LARSON, BOSWELL, KANOLD, STIFF

Passport
to **Algebra** and **Geometry**

Formal Assessment

by Robyn Silbey and
Rita M. Browning

Formal Assessment includes a diagnostic test; short quizzes following every second lesson; mid-chapter tests in Forms A and B; chapter tests in Forms A (standard), B (multiple choice), and C (more difficult); and cumulative tests following every third chapter.

McDougal Littell
A HOUGHTON MIFFLIN COMPANY
Evanston, Illinois • Boston • Dallas

Diagnostic Test of Toolbox Skills
The Diagnostic Test (pages D1–D11 with answers on pages D12 and D13) covers the 20 skills reviewed in the Toolbox of *Passport to Algebra and Geometry* (pages 742–762).

Short Quizzes
The Short Quizzes cover two lessons and are of average difficulty.

Mid-Chapter Tests
For each chapter in the Student Text, there are two Mid-Chapter Tests, Forms A and B, which are of average difficulty.

Chapter Tests
There are three Chapter Tests for each chapter in the Student Text. Forms A and B are of average difficulty with Form B in multiple choice format. Form C is more challenging.

Cumulative Tests
Cumulative Tests are of average difficulty and are provided after every third chapter in the Student Text.

Ch.	Test	Use After	Form A Page	Form B Page	Form C Page
1	Short Quiz	1.2	1		
	Short Quiz	1.4	2		
	Mid-Chapter Test	1.4	3	4	
	Short Quiz	1.6	5		
	Short Quiz	1.8	6		
	Chapter Test	Chapter 1	7	10	13
2	Short Quiz	2.2	16		
	Short Quiz	2.4	17		
	Mid-Chapter Test	2.4	18	19	
	Short Quiz	2.6	20		
	Short Quiz	2.8	21		
	Chapter Test	Chapter 2	22	25	28
3	Short Quiz	3.2	31		
	Short Quiz	3.4	32		
	Mid-Chapter Test	3.4	33	34	
	Short Quiz	3.6	35		
	Short Quiz	3.8	36		
	Chapter Test	Chapter 3	37	40	43
1–3	Cumulative Test	Chapter 3	46		
4	Short Quiz	4.2	54		
	Short Quiz	4.4	55		
	Mid-Chapter Test	4.4	56	57	
	Short Quiz	4.6	58		
	Short Quiz	4.8	59		
	Chapter Test	Chapter 4	60	63	66
5	Short Quiz	5.2	69		
	Short Quiz	5.4	70		
	Mid-Chapter Test	5.4	71	72	
	Short Quiz	5.6	73		
	Short Quiz	5.8	74		
	Chapter Test	Chapter 5	75	78	81
6	Short Quiz	6.2	84		
	Short Quiz	6.4	85		
	Mid-Chapter Test	6.5	86	87	
	Short-Quiz	6.6	88		
	Short Quiz	6.8	89		
	Chapter Test	Chapter 6	90	93	96
1–6	Cumulative Test	Chapter 6	99		

In Exercises 1 and 2, use the problem solving plan to solve each problem. A problem may contain information that is not needed or it may not have enough information to be solved. If there is not enough information, write *cannot be solved*. (Toolbox, pp. 742–743)

1. Paula Rodriguez ordered some bead necklaces for $11 each. At a craft fair she sold all but two of them for $20 each. If her profit on the necklaces was $140, how many necklaces did she order?

   1. _____

2. A science test worth 20 points consisted of two parts. Each question in the first part was worth three points and each question in the second part was worth four points. Michael Yee's score was 17. How many questions in each part did he answer correctly?

   2. _____

In Exercises 3 and 4, choose an appropriate strategy and solve the problem. (Toolbox, pp. 744–745)

3. Suppose you have 24 ft of fencing to use for a rectangular garden. Find the dimensions of the garden with the greatest possible area that can be created with 24 ft of fencing.

   3. _____

4. The diagram below shows the following pattern: one row with one equilateral triangle; two rows with four equilateral triangles, three rows with nine equilateral triangles. If you extended the diagram, how many small equilateral triangles would be in a large triangle with 20 rows?

   4. _____

In Exercises 5–8, use fraction concepts to find the answers. (Toolbox, p. 746)

5. Write $\frac{15}{18}$ in simplest form.

5. _____

6. Find $\frac{3}{4}$ of 32.

6. _____

What fraction of a dollar is 60 cents?

7. _____

8. Write $\frac{9}{20}$

  a. as a decimal   **b.** as a percent

8a._____

8b._____

In Exercises 9–12, add, subtract, multiply, or divide. If possible, simplify the result. (Toolbox, p. 747)

9. **a.** $\quad + \frac{7}{12}$ **b.** $\frac{2}{3} + \frac{3}{4}$

9a._____

9b._____

10. **a.** $\frac{5}{9}$ **b.** $\frac{7}{10} - \frac{1}{6}$

10a._____

10b._____

11. **a.** $\frac{3}{4} \cdot \frac{6}{7}$ **b.** $\frac{2}{15} \cdot \frac{3}{8}$

11a._____

11b._____

12. **a.** $18 \div \frac{2}{3}$ **b.** $\frac{9}{10} \div \frac{3}{5}$

12a._____

12b._____

Passport to Algebra and Geometry

In Exercises 13–16, add, subtract, multiply, or divide the integers.
(Toolbox, p. 748)

13. a. $-9 + 2$ b. $-6 + (-6)$ 13a. _____

                                                              13b. _____

14. a. $7 - (-1)$ b. $-3 - 2$ 14a. _____

                                                                14b. _____

15. a. $-2 \cdot 5$ b. $(-7)(-8)$ 15a. _____

                                                                15b. _____

16. a. $-24 \div (-8)$ b. $-\frac{81}{9}$ 16a. _____

                                                                16b. _____

In Exercises 17 and 18, find the factors or the prime factorization.
(Toolbox, p. 749)

17. Find the factors of 75.   17. _____

18. Find the prime factorization of 80.   18. _____

In Exercises 19 and 20, find the least common multiple of each pair of
numbers. (Toolbox, p. 749)

19. 18 and 30   19. _____

20. 42 and 55   20. _____

In Exercises 21 and 22, evaluate the exponent or write the number as a power of 10. (Toolbox, p. 750)

21. Evaluate each exponent.

    a. $(-2)^5$ b. 10^{-2}

21a._____

21b._____

22. Write each number as a power of 10.

    a. 1000   b. 0.001

22a._____

22b._____

In Exercises 23 and 24, write each number in scientific notation. (Toolbox, p. 750)

23. 0.00008

23._____

24. 9,170,000

24._____

Exercises 25–28, solve the equation and check the solution.
olbox, p. 751)

    a. $x + 7 = 15$ b. $-8 + t = -2$

25a._____

25b._____

26. a. $m - (-2) = 2$ b. $n - 5 = -8$

26a._____

26b._____

27. a. $4y = 32$ **b.** $-3z = -30$

27a. _____

27b. _____

28. a. $w \div 2 = -8$ **b.** $\dfrac{a}{-6} = -3$

28a. _____

28b. _____

In Exercises 29–32, solve each proportion. (Toolbox, p. 752)

29. $\dfrac{x}{32} = \dfrac{15}{48}$

29. _____

30. $\dfrac{54}{k} = \dfrac{90}{72}$

30. _____

31. $\dfrac{75}{105} = \dfrac{r}{28}$

31. _____

32. $\dfrac{2}{0.65} = \dfrac{3.6}{n}$

32. _____

In Exercises 33–36, solve the percent problem. (Toolbox, p. 753)

33. What is 36% of 625?

33. _____

34. What percent of 12 is 21?

34. _____

35. 57 is $66\frac{2}{3}\%$ of what number?

35. _____

36. 4% of what number is 6?

36. _____

In Exercises 37–40, find the area of each figure. (Toolbox, p. 754)

37. a rectangle that has a length of 8.8 cm and a width of 5 cm

37. _____

38. a triangle with a base of 24 in. and a height of 9 in.

38. _____

39.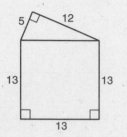

39. _____

40.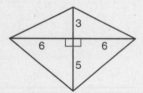

40. _____

In Exercise 41, evaluate each square root. If necessary, round the result to two decimal places. (Toolbox, p. 755)

41. a. $\sqrt{64}$ **b.** $\sqrt{132}$

41a. _____

41b. _____

In Exercises 42–44, a and b are the lengths of the legs of a right triangle, and c is the length of the hypotenuse. Find the unknown length. If necessary, round the result to two decimal places. (Toolbox, p. 755)

42. $a = 20, b = 21$

42. _____

43. $a = 10, c = 14$

43. _____

44. $b = 6, c = 8$

44. _____

In Exercises 45–48, find the surface area and the volume of each cube and rectangular prism. (Toolbox, p. 756)

45. a cube with edges of length 1.2 m

45. _____

46. a cube with edges of length $\frac{3}{4}$ ft

46. _____

47. a rectangular prism with length 10 cm, width 7 cm, and height 4 cm

47. _____

48. a rectangular prism with length 8 in., width 6 in., and height $2\frac{1}{2}$ in.

48. _____

In Exercises 49–52, classify the triangle according to its sides (*scalene, isosceles, equilateral*) and its angles (*acute, right, obtuse*). (Toolbox, p. 757)

49.

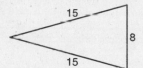

49. _____

_____

50.

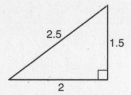

50. _____

_____

51.

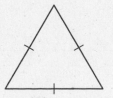

51. _____

_____

52.

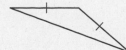

52. _____

_____

In Exercises 53 and 54, use the diagram below. (Toolbox, p. 758)

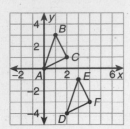

53. Describe the translation of triangle *ABC* to triangle *DEF*.

53. _____

54. Find the coordinates of triangle *ABC* when it is translated 4 units up.

54. _____

In Exercises 55 and 56, tell whether the diagram shows a translation of one figure to another. Write *yes* or *no*. (Toolbox, p. 758)

55.

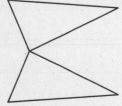

55. _____

56.

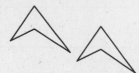

56. _____

In Exercises 57–60, use counting methods to solve the problem. (Toolbox, p. 759)

57. At a banquet dinner you can choose fish, chicken, or vegetarian lasagna as an entree, and either apple pie or ice cream for dessert. Find the number of different dinners that can be ordered.

57. _____

In Exercises 69 and 70, use the following information. In a recent year, 39% of the beds Americans bought were twin beds, 26% were full-size, 24% were queen-size, 6% were king-size, and 5% were custom-made. (Toolbox, p. 762)

69. Draw a circle graph to represent the data.   69.

70. Use your circle graph from Exercise 69 to explain how you know   70.
    that about half of the beds bought by Americans in a recent year
    were full-size or queen-size.

In Exercises 71 and 72, use the table showing the number of wins earned by the 15 teams in the American Conference of the National Football League during the 1996 season. *(Source: 1996 World Book Yearbook)* (Toolbox, p. 762)

Number of wins	0–3	4–7	8–11	12–15
Number of teams	1	3	10	1

71. Draw a circle graph to display the data.   71.

72. Explain how you can use the circle graph to estimate the fraction of   72.
    the teams that won between 8 and 11 games.

Answers to Diagnostic Test of Toolbox Skills

1. 20 necklaces

2. 3 correct answers in first part; 2 correct answers in the second part

3. 6 ft by 6 ft

4. 400 small equilateral triangles

5. $\frac{5}{6}$ **6.** 24   **7.** $\frac{3}{5}$

8. a. 0.45   **b.** 45%

9. a. $\frac{2}{3}$ **b.** $\frac{17}{12}$, or $1\frac{5}{12}$

10. a. $\frac{2}{9}$ **b.** $\frac{8}{15}$

11. a. $\frac{9}{14}$ **b.** $\frac{1}{20}$

12. a. 27   **b.** $\frac{3}{2}$, or $1\frac{1}{2}$

13. a. -7 **b.** -12

14. a. 8   **b.** -5

15. a. -10 **b.** 56

16. a. 3   **b.** -9

17. 1, 3, 5, 15, 25, 75   **18.** $2^4 \cdot 5$

19. 90   **20.** 2310

21. a. -32 **b.** $\frac{1}{100}$, or 0.01

22. a. 10^3 **b.** 10^{-3}

23. 8×10^{-5} **24.** 9.17×10^6

25. a. 8   **b.** 6

26. a. 0   **b.** -3

27. a. 8   **b.** 10

28. a. -16 **b.** 18

29. 10   **30.** 43.2   **31.** 20   **32.** 1.17   **33.** 225

34. 175%   **35** 85.5   **36.** 150   **37.** 44 cm^2

38. 108 in^2 **39.** 199 sq. units   **40.** 48 sq. units

41. a. 8   **b.** 11.49

42. $c = 29$ **43.** $b \approx 9.80$ **44.** $a \approx 5.29$

45. 8.64 m^2; 1.728 m^3 **46.** $\frac{27}{8}$ ft^2; $\frac{27}{64}$ ft^3

47. 276 cm^2; 280 cm^3 **48.** 166 in.^{2}; 120 in.^{3}

49. isosceles, acute   **50.** scalene, right

51. equilateral, acute   **52.** isosceles, obtuse

53. 2 units to the right and 4 units down

54. $(0, 4)$, $(1, 7)$, $(2, 5)$ **55.** no   **56.** yes

57. 6   **58.** 676   **59.** 8   **60.** 240

61. permutation; $4! = 24$

62. combination; 15

63. combination; 36

64. permutation; 6

65.

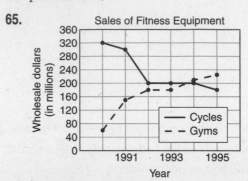

66. *Sample answer:* From 1990 to 1995, the sales of home gyms has more than tripled, while the sales of stationary cycles has decreased by more than 40 percent.

67.

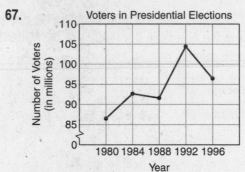

68. *Sample answer:* Voter turnout was highest for the 1992 Presidential election.

69.

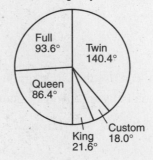

70. About half the circle consists of the parts labeled "full-size" and "queen-size".

71. Wins Earned by 15 Teams

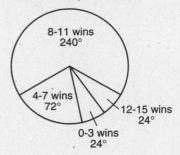

8-11 wins
240°

4-7 wins
72°

12-15 wins
24°

0-3 wins
24°

72. Look at what part of the circle the "8–11" portion covers. It is about two thirds of the circle, so about $\frac{2}{3}$ of the teams won 8–11 games.

1.2 Short Quiz

Name_____

Date _____

In Exercises 1 and 2, describe the pattern. Then list the next 3 numbers.

1. 6, 12, 18, 24, ⬚?⬚ , ⬚?⬚ , ⬚?⬚

1. _____

2. 54, 48, 42, 36, ⬚?⬚ , ⬚?⬚ , ⬚?⬚

2. _____

3. You run the first 5K of a 10K footrace in 29:29.9, and the second 5K in 30:00. During which half of the race was your time faster? Explain.

3. _____

4. Write a verbal description for the number sentence 35 + 479 = 514.

4. _____

5. Find the difference of 923.4 and 375.8.

5. _____

In Exercises 6 and 7, find the product or quotient.

6. (8.3)(7.9)

6. _____

7. $\dfrac{2187}{81}$

7. _____

8. Make up your own pattern using 7 numbers. Then describe it.

8. _____

Name_____

Date _____

In Exercises 1 and 2, write a verbal description for the number sentences.

1. $8^3 = 512$

1._____

2. $\sqrt{1.44} = 1.2$

2._____

3. Write $4 \times 4 \times 4 \times 4 \times 4$ as a power. Use a calculator to find the value of the power.

3._____

4. Use a calculator to find the value of the expression $\sqrt{1.02}$. Round your result to two decimal places.

4._____

5. Evaluate the expression $5[12 - (5 + 1) \div 3]$.

5._____

6. Insert parentheses to make the expression $8 \cdot 2 + 3 \div 4 = 10$ true.

6._____

7. Write a numerical expression for the phrase 27 divided by the difference of 42 and 39. Then evaluate the expression.

7._____

8. Insert parentheses in $9 \cdot 8 - 6 \div 2 + 4$ to create two expressions with different values. Use < or > to compare the expressions.

8.

Name_____

Date_____

In Exercises 1 and 2, describe the pattern. Then list the next 3 numbers. (1.1)

1. 4, 8, 12, $\boxed{?}$, $\boxed{?}$, $\boxed{?}$

1. _____

2. $\frac{1}{2}, \frac{1}{3}, \frac{1}{4}, \boxed{?} , \boxed{?} , \boxed{?}$

2. _____

In Exercises 3 and 4, write a verbal description of the expression and evaluate the expression. (1.2, 1.3)

3. 6^3

3. _____

4. $\dfrac{203}{29}$

4. _____

In Exercises 5–7, imagine that you are at an amusement park. Ground-vehicle rides cost 7 tickets each and water rides cost 9 tickets each. You plan to go on two of each kind of ride. (1.4)

5. How many tickets will you need?

5. _____

6. You buy a 50-ticket package. After completing your rides and spending an additional 12 tickets on skee ball, how many tickets are left?

6. _____

7. Choose a combination of ground-vehicle and water rides that will use a total of between 45 and 50 tickets. On how many of each type of ride will you go? How many tickets will you use?

7. _____

In Exercises 8 and 9, consider a square picture frame with an area of 256 square inches. (1.4)

8. What is the length of each side of the picture frame?

8. _____

9. Create a combination of 4-by-4 or 8-by-8 pictures that will fit perfectly inside your picture frame. How many of each size picture did you use?

9. _____

In Exercises 1 and 2, describe the pattern. Then list the next 3 numbers. (1.1)

1. 7, 14, 21, ?, ?, ?

1. _____

2. $\frac{1}{2}$, 1, $\frac{3}{2}$, ?, ?, ?

2. _____

In Exercises 3 and 4, write a verbal description of the expression and evaluate the expression. (1.2, 1.3)

3. 4^4

3. _____

4. $\frac{243}{27}$

4. _____

In Exercises 5–7, imagine that you are at a video arcade. Interactive games cost 6 tokens each and seated games cost 4 tokens each. You plan to play 3 of each type of game. (1.4)

5. How many tokens will you need?

5. _____

6. You get 50 tokens for $10.00. After completing your games and spending 4 more tokens on pinball games, how many tokens are left?

6. _____

7. Choose a combination of interactive games and seated games that will use a total of between 45 and 50 tokens. How many of each type of game will you play? How many tokens will you use?

7. _____

In Exercises 8 and 9, consider a square quilt with an area of 144 square feet. (1.4)

8. What is the length of each side of the quilt?

8. _____

9. Create a combination of 3-by-3 or 6-by-6 pieces that could be used to make the quilt. How many of each size piece did you use?

9. _____

Name_____

Date _____

1. Evaluate $a(3 + a)$ for $a = 5$.

1._____

2. Evaluate $x^2 + (y + 5)$ for $x = 2$ and $y = 9$.

2._____

3. Write the algebraic expression for *the sum of a number and 3*.

3._____

4. Write the verbal description for $t \div 8$.

4._____

5. If the expression $2x + 6$ has a value of 20, what is the value of x?

5._____

For Exercises 6–9, use the double bar graph showing club memberships at Gaithersburg Middle School.

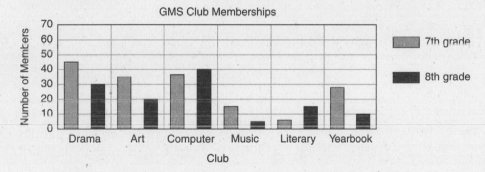

6. About how many 7th grade students belong to the drama club?

6._____

7. In what club is the difference between the 7th and 8th grade students the greatest? the least?

7._____

8. The school wants to merge two of the smaller clubs to form a single club with between 50 and 75 students. Which two clubs do you suggest combining? Explain.

8._____

9. What will be the estimated total membership when the two clubs are combined?

9._____

1.8 Short Quiz

Name_____

Date _____

1. Draw a pentagon with four sides equal in length and the fifth side a different length.

   1.

2. Draw a diagram showing the total number of diagonals in an octagon.

   2.

3. Find the area of a triangle whose base is 14 feet and height is 6 feet. (Use the formula $\frac{1}{2}$(base · height).)

   3._____

In Exercises 4 and 5, decide whether the figure is a polygon. If it is, name it. If it is not, explain why.

4.

   4.

5.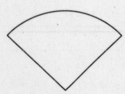

   5.

6. Use a calculator to evaluate the expressions below. Then describe the pattern.

   10×12
   11×13
   12×14
   13×15
   14×16

   6.

7. Make up a pattern of your own using the calculator. Write the first six numbers in the pattern. Then describe it.

   7.

In Exercises 1 and 2, describe the pattern. Then list the next 3 numbers. (1.1)

1. 8, 16, 24, 32

1._____

2. 108, 97, 86, 75

2._____

In Exercises 3–6, evaluate the expression. (1.2, 1.4)

3. $(4.79)(12.3)$

3._____

4. $2496 \div 6$

4._____

5. $4 \cdot 7 + 45 \div 15$

5._____

6. $[(6 + 8) \div 2] + 63 \div 21$

6._____

In Exercises 7 and 8, write the expression as a power. Then evaluate. (1.3)

7. $4 \cdot 4 \cdot 4 \cdot 4 \cdot 4$

7._____

8. $\frac{1}{2} \cdot \frac{1}{2} \cdot \frac{1}{2} \cdot \frac{1}{2} \cdot \frac{1}{2} \cdot \frac{1}{2}$

8._____

In Exercises 9 and 10, use a calculator to evaluate the expression. Round to 2 decimal places. (1.3)

9. $\sqrt{12.25}$

9._____

10. $\sqrt{410}$

10._____

11. A square hand-stitched carpet has an area of 529 square inches. What are the lengths of the sides of the carpet? (1.4)

11._____

In Exercises 12–14, evaluate the expression for $k = 9$ and $s = 6$. (1.3–1.5)

12. $3k - 2s$

12._____

13. $\frac{1}{2}(4k + 3s) \div 3$

13._____

14. $s^2 + 3\sqrt{k}$

14._____

15. Evaluate $4(h^2 + 3w) - 5dh$ for $h = 3$, $w = 4$, and $d = 5$. (1.3–1.5)

15._____

In Exercises 16–18, use the bar graph below. *(Source: 1997 World Almanac)* **(1.6)**

National Hockey League, 1995-1996
Central Division

16. About how many games did Toronto win?

16._____

17. Which team had the smallest difference between wins and losses? The greatest difference?

17._____

18. Write a question that can be answered by interpreting the graph. Then write the solution.

18.

In Exercises 19–21, decide whether the figure is a polygon. If it is, name it. (1.7)

19.

19._____

20.

20._____

21.

21._____

Statistics for some major U.S. Zoological parks are listed in the table. Use the table to answer Exercises 22 and 23. *(Source: 1997 World Almanac)* **(1.6)**

Zoo/Location	Budget (in millions)	Attendance (in millions)	Acres	Species
Audubon, LA	10.0	0.9	58	360
Cleveland, OH	8.7	1.2	165	574
Dallas, TX	6.9	0.4	70	377
Lincoln Park, IL	12.0	2.0	35	233
San Francisco, CA	12.0	0.9	125	270

22. Which of the parks listed has the highest annual attendance?

22._____

23. Which parks have over 300 different species?

23._____

In Exercises 24–26, imagine that you can ride your bike downhill at a rate of about 12 miles per hour. Your rate uphill is about 6 miles per hour. (1.5)

24. How many minutes will it take you to ride about 3 miles downhill?

24._____

25. How much farther can you travel downhill than uphill in $\frac{1}{2}$ hour?

25._____

26. It takes you 40 minutes to ride up a steep hill. How long will it take you to ride downhill to your starting point?

26._____

In Exercises 1 and 2, what are the next three numbers in the pattern? (1.1)

1. 6, 12, 18, ?, ?, ?

   a. 24, 30, 36   **b.** 22, 28, 32   **c.** 22, 26, 28   **d.** 26, 32, 38

   1._____

2. 95, 88, 81, ?, ?, ?

   a. 75, 68, 61   **b.** 74, 67, 60   **c.** 74, 66, 58   **d.** 75, 67, 60

   2._____

In Exercises 3–8, what is the value of the expression? (1.2–1.4)

3. $(5.24)(38.6)$

   a. 202,264   **b.** 2022.64   **c.** 202.264   **d.** 20.2264

   3._____

4. $4088 \div 28$

   a. 146   **b.** 140.96   **c.** 146.03   **d.** 141

   4._____

5. $8 \cdot 6 + 36 \div 4$

   a. 21   **b.** 12.5   **c.** 30.5   **d.** 57

   5._____

6. $(14 - 6) \cdot 5 - 7 + 36 \div 6$

   a. 26   **b.** 11.5   **c.** 39   **d.** 33

   6._____

7. 5^4

   a. 3125   **b.** 20   **c.** 125   **d.** 625

   7._____

8. $(4.2)^3$

   a. 74.088   **b.** 12.6   **c.** 17.64   **d.** $(4.2)(4.2)(4.2)$

   8._____

In Exercises 9 and 10, what is the value of the expression, rounded to 2 decimal places? You may use a calculator to help you. (1.3)

9. $\sqrt{32}$

   a. 5.66   **b.** 56.6   **c.** 5.6   **d.** 5.65

   9._____

10. $\sqrt{2.25}$

    a. 15   **b.** 1.125   **c.** 1.5   **d.** 5.0625

    10._____

In Exercises 11 and 12, what is the value of the expression when *j* = 7 and *k* = 4? (1.4, 1.5)

11. $5j - 6k$

    a. 11   **b.** 18   **c.** 59   **d.** 9

11._____

12. $\frac{1}{3}(2j + 4k)$

    a. 7.333   **b.** 10   **c.** 30   **d.** 74.666

12._____

The number of cars sold by type, per 100, is listed in the table. Use the table to answer Exercises 13–15. (1.6)

U.S. Car Sales by Vehicle Size and Type, 1991–1995 *(Source: 1997 World Almanac)*

Year	Small	Midsize	Large	Luxury	Total
1991	33.0	44.9	8.3	13.9	100.0
1992	32.9	44.5	9.2	13.4	100.0
1993	32.8	43.3	11.1	12.8	100.0
1994	29.2	45.6	11.7	13.5	100.0
1995	27.1	48.5	10.8	13.6	100.0

13. In what year were the most midsize cars sold?

    a. 1994   **b.** 1991   **c.** 1995   **d.** 1992

13._____

14. How many more large cars per 100 were sold in 1992 than in 1991?

    a. 0.09   **b.** 17.5   **c.** 9.0   **d.** 0.9

14._____

15. Which type of car would you predict a decrease in sales for 1996?

    a. small   **b.** midsize   **c.** large   **d.** luxury

15._____

16. A square coffee table has an area of 441 square inches. What are the lengths of the sides of the table? (1.4)

    a. 110.25 in.   **b.** 20 in.   **c.** 21 in.   **d.** 111 in.

16._____

In Exercises 17 and 18, what is the name of the polygon shown? (1.7)

17.

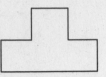

    a. regular pentagon   **b.** hexagon

    c. regular nonagon   **d.** octagon

17._____

18.

    a. regular heptagon   **b.** decagon

    c. quadrilateral   **d.** pentagon

18._____

19. Which of the figures is a regular octagon? (1.7)

    a.   **b.**   **c.**   **d.**

19._____

In Exercises 20 and 21, use the following information.
At Aviation School, regular classrooms have tables and chairs for 32
students. Special labs have facilities for 16 students. (1.5)

20. Which of the statements below is the algebraic expression represent-
ing the number of students that would fit in 12 regular classrooms and
a special lab?

    a. $12 \cdot 32 \cdot 16$ **b.** $(12 \cdot 32) + 16$ **c.** $12 \cdot (32 + 16)$ **d.** $12 + 32 + 16$

20._____

21. What is the total number of students, if all seats in all classrooms are
filled?

    a. 60   **b.** 6144   **c.** 400   **d.** 576

21._____

In Exercises 22 and 23, use the formula $d = r \cdot t$ (Distance = Rate · Time). (1.5)

22. If a car travels at 55 miles per hour for 3.5 hours, what is its total distance
traveled?

    a. 1925 miles   **b.** 165 miles

    c. 192.5 miles   **d.** 220 miles

22._____

23. A car travels 393 miles in 7.5 hours. What is its rate?

    a. 56 miles per hour   **b.** 29.475 miles per hour

    c. 2947.5 miles per hour   **d.** 52.4 miles per hour

23._____

In Exercises 1 and 2, describe the pattern. Then list the next 3 numbers. (1.1)

1. 2, 8, 32, 128   1. _____

2. 100, 99, 97, 94   2. _____

In Exercises 3 and 4, write the first 6 numbers in the sequence. (1.1, 1.3)

3. The first two numbers are 1 and 2. Each succeeding number is the   3. _____
   sum of the two preceding numbers.

4. The first two numbers are 1.41 and 2. Each succeeding number is   4. _____
   the square root of the next even number. Round numbers to two
   decimal places.

In Exercises 5–8, evaluate the expression. (1.2–1.4)

5. $0.0325 - 0.00987$ 5. _____

6. $111.54 \div 21.45$ 6. _____

7. $53 - (4 + 7)(20 - 19)$ 7. _____

8. $(23 + 33) \div \sqrt{25}$ 8. _____

In Exercises 9 and 10, write the expression as a power. Then evaluate. (1.3)

9. $4.5 \cdot 4.5 \cdot 4.5 \cdot 4.5$ 9. _____

10. $\frac{4}{5} \cdot \frac{4}{5} \cdot \frac{4}{5} \cdot \frac{4}{5}$ 10. _____

11. A rectangular calendar has an area of 338 square inches. When folded   11. _____
    in half, it forms a square. What is the length of the longer side? (1.4)

12. Calculate the product of 246 and the first 5 even numbers. Record the results in a table. Describe the pattern. (1.1, 1.8)

12.

In Exercises 13–16, evaluate the expression $r = 5$, $s = 9$, **and** $t = 12$. (1.3–1.5)

13. $r \cdot s + t^2$

13._____

14. $\frac{1}{4}(r \cdot t) - s$

14._____

15. $6(s - 1)^2$

15._____

16. $\sqrt{3t} + \dfrac{s}{3}$

16._____

In Exercises 17–19, use the double bar graph below. (1.6)

(Source: 1997 World Almanac)

Education, Gender, and Monthly Income, 1994

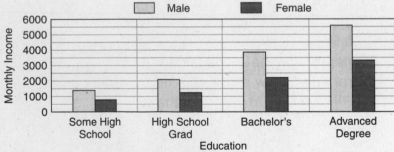

17. How much more does a male earn with an advanced degree than with a bachelor's degree?

17._____

18. At which level of education is the difference greatest in monthly income between males and females?

18._____

19. What generalizations can you make about education, gender, and monthly income from interpreting the graph?

19.

20. Explain what is incorrect about the solution process. Show the correct steps. Write the solution. (1.4)

$$6 + 4 \cdot (8 - 5)^3 - 20 \div 2$$
$$10 \cdot (8 - 5)^3 - 20 \div 2$$
$$10 \cdot (3)^3 - 20 \div 2$$
$$10 \cdot 27 - 20 \div 2$$
$$270 - 20 \div 2$$
$$250 \div 2$$
$$125$$

20. _____

In Exercises 21 and 22, draw and describe the named polygon. (1.7)

21. Regular pentagon

21. _____

22. Octagon

22. _____

For Exercises 23–26, imagine that you are preparing for a 10-mile walk/run footrace. You can run at a speed of about 6 miles per hour, and you can walk at a speed of about 4 miles per hour. (1.5)

23. Calculate the difference in time if you were running the entire course instead of walking the entire course.

23. _____

24. You ran for 8 miles and walked the rest of the way. Write an algebraic model that represents the total time it took you to complete the race.

24. _____

25. How long did it take to complete the race?

25. _____

26. Construct your own run/walk program for the footrace. Write an algebraic model to find the time it would take you to complete the race. Then solve.

26. _____

Name_____

Date _____

1. Use the Distributive Property to write an equivalent expression for $3(x + 2)$.

1. _____

2. Illustrate the result of Exercise 1 with an algebra tile sketch.

2.

3. Use the Distributive Property to rewrite $15(x + 4)$.

3. _____

4. Simplify the expression $8x + 3 + 4x$.

4. _____

Simplify the expression $3(b + 3) + 3 + 3b$.

5. _____

6. Write an expression that has three terms and simplifies to $10p + 6$.

6. _____

7. Write an expression for the perimeter of the hexagon below.

7. _____

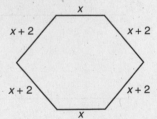

8. Find the perimeter of the hexagon above when $x = 6$.

8. _____

2.4 Short Quiz

Name_____

Date _____

In Exercises 1–3, simplify the expression.

1. $5a + 3 + 6a$

1. _____

2. $4a^2x + 17ax + 3a^2x$

2. _____

3. $12(w - r) + 3(w - r)$

3. _____

4. Write an expression that has 3 terms and simplifies to $10x + 4y$.

4. _____

5. Evaluate the expression $8gh + 3g$ when $g = 6$ and $h = 9$.

5. _____

In Exercises 6 and 7, write the question as an equation. Then solve it mentally.

6. What number can be multiplied by 9 to obtain 72?

6. _____

7. What number can be subtracted from 85 to obtain 76?

7. _____

In Exercises 8 and 9, solve the equation. Then write another equation that has the same solution.

8. $r + 14 = 21$

8. _____

9. $\dfrac{e}{5} = 3$

9. _____

Form A
(Use after Lesson 2.4)

Name_____

Date _____

In Exercises 1–4, rewrite and/or simplify the expression. (2.1, 2.2)

1. $8(a + 4)$

1. _____

2. $5(x + 3y + 5)$

2. _____

3. $7c + 3 + 6c$

3. _____

4. $2(x + 7) + 9x$

4. _____

5. Evaluate the expression $4(3h + 7)$ for $h = 6$. (2.2)

5. _____

In Exercises 6 and 7, use the rectangle below. (2.1, 2.2)

6. Write a simplified expression for the perimeter of the rectangle.

6. _____

7. Write a simplified expression for the area of the rectangle.

7. _____

In Exercises 8–10, solve the equation. (2.3, 2.4)

8. $83 = p + 45$

8. _____

9. $v - 8 = 27$

9. _____

10. $71 + b = 110$

10. _____

11. Draw and label a triangle that has the same perimeter as the polygon at the right. (2.2–2.4)

11.

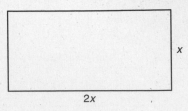

2x

Name _____

Date _____

In Exercises 1–4, rewrite and/or simplify the expression. (2.1, 2.2)

1. $7(t + 5)$

1. _____

2. $3(i + 8 + 4j)$

2. _____

3. $2u + 15 + 12u$

3. _____

4. $5(b + 9) + 2b$

4. _____

5. Evaluate the expression $7(8g + 4)$ for $g = 5$. (2.2)

5. _____

In Exercises 6 and 7, use the rectangle below. (2.1, 2.2)

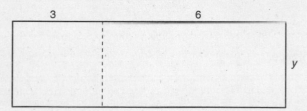

6. Write a simplified expression for the perimeter of the rectangle.

6. _____

7. Write a simplified expression for the area of the rectangle.

7. _____

In Exercises 8–10, solve the equation. (2.3, 2.4)

8. $75 = 91 - a$

8. _____

9. $56 + k = 80$

9. _____

10. $5 + y = 42$

10. _____

11. Draw and label a rectangle that has the same perimeter as the polygon at the right. (2.2–2.4)

11.

2.6 Short Quiz

Name_____

Date _____

In Exercises 1–3, solve the equation.

1. $5y = 65$

2. $\dfrac{v}{6} = 16$

3. $3f = 4.8$

4. Draw a rectangle that has a length of 9 units and an area of 36 square units. Find the width.

In Exercises 5 and 6, write an equation that represents the sentence. Then solve.

5. The number of cars c times 4 tires is 24 tires.

6. The number of video games v divided by 6 equals 3 video games.

7. Write a real life situation that can be solved using the equation $2f = 10$.

In Exercises 8 and 9, translate the verbal phrase into an algebraic expression.

8. The difference of 3 times a number and 5

9. The product of 7 and a number, increased by 2

1._____

2._____

3._____

4._____

5._____

6._____

7.

8._____

9._____

2.8 Short Quiz

Name _____

Date _____

In Exercises 1 and 2, write an algebraic equation that represents the verbal sentence. Then solve the equation.

1. The number of tennis balls, t, increased by 6 is 15.

   1. _____

2. The number of new, unsharpened pencils, w, decreased by 43 equals 8 pencils left to be sharpened.

   2. _____

3. Write a verbal sentence that represents the equation $16 = c + 5$. Then solve the equation.

   3.

4. One number is 398 more than another number. The larger number is 500. What is the smaller number? Write an algebraic model to answer the question. Then solve the equation.

   4.

In Exercises 5–8, imagine that you are saving for a video game that costs $47.25 (including tax). You save $12.50 in January, $9.75 in February, and $15.75 in March. How much more must you save before you can afford the video game?

5. Write a verbal model that relates the total cost, the amount saved thus far, and the amount yet to be saved.

   5.

6. Assign labels to the three parts of your model.

   6.

7. Use the labels to translate your verbal model into an algebraic model.

   7. _____

8. Solve the algebraic model.

   8. _____

For Exercises 1 and 2, use the Distributive Property to rewrite the expression. (2.1)

1. $(3 + 4)$

1._____

2. $8(c + 2)$

2._____

In Exercises 3–6, simplify the expression. (2.1, 2.2)

3. $4p + 5p$

3._____

4. $x^2 + 3x^2 + 5x$

4._____

5. $2y + 5 + 7 + 4y$

5._____

6. $5(b + 3) + 2b + 7$

6._____

In Exercises 7–9, use mental math to solve the equation. (2.3)

7. $28 - n = 17$

7._____

8. $6q = 42$

8._____

9. $16r = 64$

9._____

In Exercises 10–13, write an algebraic model for the phrase or sentence. (2.6, 2.7)

10. The sum of a number and 46

10._____

11. Twenty-eight students receive an equal share.

11._____

12. The difference of a number and 23 is 18.

12._____

13. The product of 9 and a number is 63.

13._____

In Exercises 14–17, solve the equation. (2.3–2.5)

14. $23 + f = 61$　　　　　　　　　　　　　**14.**_____

15. $92 = y - 67$　　　　　　　　　　　　　**15.**_____

16. $1.5g = 13.5$　　　　　　　　　　　　　**16.**_____

17. $\dfrac{a}{3} = 16$　　　　　　　　　　　　　**17.**_____

In Exercises 18–21, solve the inequality. (2.9)

18. $y - 6 \le 35$　　　　　　　　　　　　　**18.**_____

19. $8p > 68$　　　　　　　　　　　　　　**19.**_____

20. $9.5 + m \ge 100$　　　　　　　　　　　**20.**_____

21. $\dfrac{h}{4} \le 12$　　　　　　　　　　　　　**21.**_____

In Exercises 22–24, write an algebraic equation or inequality and solve. (2.7, 2.9)

22. The amount of money in your bank decreased by $10.50 is $27.75.　　**22.**_____

23. $2.50 per week times x weeks is at least $15.00.　　**23.**_____

24. 215 minutes divided by 5 days is the number of minutes in math class per day.　　**24.**_____

For Exercises 25–28, use the region at the right. (2.6, 2.7)

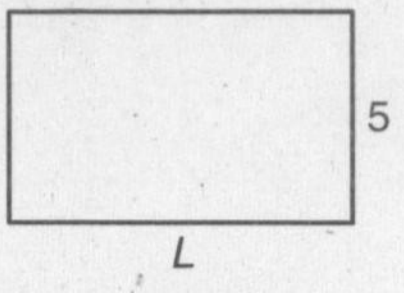

[rectangle with side labeled 5 and bottom labeled L]

25. Write an expression for the perimeter of the region.

25._____

26. Write an expression for the area of the region.

26._____

Find the perimeter and area of the region if $L = 8$.

27._____

28. Make up another length for L between 7 and 8. Find the perimeter and area of the region using your length.

28._____

For Exercises 29–33, use the following information.

The original stars and stripes flag design, with 13 stars, was first flown on U.S. soil in 1777. Stars were added as states entered the Union, and by 1912, there were 48 stars on the American flag. Forty-eight years later, the last two stars were added to construct the 50-star flag we use today.
(Source: Crouthers, Flags of American History, c 1962) **(2.8)**

29. How many years ago was the first flag flown on U.S. soil?

29._____

30. In what year were the last two stars added to the flag?

30._____

31. Using the information given above, write a question about American flags that can be represented by an algebraic equation.

31._____

32. Write an algebraic equation that can be used to answer the question you wrote in Exercise 31.

32._____

33. Solve the equation you wrote in Exercise 32.

33._____

In Exercises 1 and 2, which expression is obtained by using the Distributive Property? (2.1)

1. $3(x + 7)$

   a. $3x + 7$ **b.** $3 + x + 10$

   c. $3x + 21$ **d.** $3x + 10$

1._____

2. $5(3 + e + 2f)$

   a. $53 + 5e + 10f$ **b.** $15 + e + 10f$

   c. $15 + 5e + 2f$ **d.** $15 + 5e + 10f$

2._____

In Exercises 3–5, simplify the given expression. (2.1, 2.2)

3. $4b + 5t + 10b$

   a. $14b + 5t$ **b.** $19 + b^2t$

   c. $9bt + 10b$ **d.** $200b^2t$

3._____

4. $4(x + 2y) + 3(y + 2x)$

   a. $4x + 8y + 3y + 6x$ **b.** $10x + 11y$

   c. $11x + 10y$ **d.** $12(3x + 3y)$

4._____

5. $3x^2 + 5x + 7 + 4x^2 + 1 + 3x$

   a. $12x^4 + 15x^2 + 7$ **b.** $12x^2 + 15x + 8$

   c. $7x^4 + 8x^2 + 8$ **d.** $7x^2 + 8x + 8$

5._____

In Exercises 6–8, which value is the solution for the given equation? (2.3–2.5)

6. $18 - m = 12$

   a. $m = 4$ **b.** $m = 30$ **c.** $m = 1.5$ **d.** $m = 6$

6._____

7. $z + 2 = 16$

   a. $z = 14$ **b.** $z = 32$ **c.** $z = 8$ **d.** $z = 18$

7._____

8. $8g = 72$

   a. $g = 80$ **b.** $g = 9$ **c.** $g = 64$ **d.** $g = 576$

8._____

In Exercises 9–13, which is the algebraic model for the sentence? (2.7, 2.8)

9. The sum of 53 and a number is 71.

   a. $53 + 71 = n$ b. $71 - 53 = n$

   c. $53 + n = 71$ d. $71 - n = 53$

   9._____

10. The difference of a number and 25 is 16.

    a. $n - 25 = 16$ b. $25 + 16 = n$

    c. $25 - 16 = n$ d. $n + 16 = 25$

    10._____

11. The product of 9 and a number is 108.

    a. $9 + 108 = n$ b. $9n = 108$

    c. $n \div 9 = 108$ d. $9 - n = 108$

    11._____

12. The quotient of 52 and a number is greater than or equal to 4.

    a. $52 \div n = 4$ b. $52 \div 4 = n$

    c. $52 \div n \geq 4$ d. $52 \div n \leq 4$

    12._____

13. The product of a number and 15 is less than 105.

    a. $15n \leq 105$ b. $15n > 105$

    c. $15n < 105$ d. $15n \geq 105$

    13._____

In Exercises 14–17, which value is the solution for the given equation? (2.3)

14. $3g + 15 = 60$

    a. $g = 45$ b. $g = 30$ c. $g = 15$ d. $g = 60$

    14._____

15. $69 = 9v - 12$

    a. $v = 9$ b. $v = 81$ c. $v = 6.33$ d. $v = 7.66$

    15._____

16. $35.4 + h = 42$

    a. $h = 6.6$ b. $h = 5.6$ c. $h = 77.4$ d. $h = 7.6$

    16._____

17. $\frac{1}{5}k = 15$

    a. $k = 3$ b. $k = 75$ c. $k = \frac{1}{3}$ d. $k = 7.5$

    17._____

In Exercises 18–21, which is the solution to the given inequality? (2.9)

18. $t + 7 > 12$

    a. $t = 5$ **b.** $t < 5$ **c.** $t > 5$ **d.** $t \geq 5$

18._____

19. $65 \leq d - 25$

    a. $d \geq 40$ **b.** $d \leq 40$ **c.** $d \geq 90$ **d.** $d \leq 90$

19._____

20. $14u \leq 168$

    a. $u \leq 12$ **b.** $u \geq 12$ **c.** $u \leq 154$ **d.** $u \leq 182$

20._____

21. $\dfrac{e}{36} \geq 4$

    a. $e \geq 9$ **b.** $e \leq 144$ **c.** $e = 9$ **d.** $e \geq 144$

21._____

In Exercises 22 and 23, imagine that you sell and deliver newspapers in your neighborhood. You earn \$0.07 per paper for delivery. You earn a \$5.00 bonus for every new subscription you obtain for the company. This week you delivered 120 papers and obtained one new subscription. (2.8)

22. Which algebraic expression tells the total amount you earned this week?

    a. $120(\$0.07) + \$5.00$ **b.** $\$5.00 + \$0.07 + \$1.20$

    c. $120(\$5.00 + \$0.07)$ **d.** $\$5.00 \cdot \$0.07 \cdot 120$

22._____

23. What were your weekly earnings?

    a. \$8.40   **b.** \$608.40   **c.** \$6.27   **d.** \$13.40

23._____

For Exercises 24 and 25, use the figure to the right. (2.6, 2.7)

24. Which simplified expression describes the perimeter?

    a. $8x$ **b.** $3x + 3x + 2x$

    c. $7x$ **d.** $2(3x) + 2x$

24._____

25. What is the perimeter of the figure if $x = 8$?

    a. $48 + 16$ **b.** 64   **c.** 54   **d.** 56

25._____

In Exercises 1 and 2, use the Distributive Property to rewrite the expression. (2.1)

1. $5(8 + 10 + 11)$ 1. _____

2. $z(3 + f + j)$ 2. _____

In Exercises 3–6, simplify the expression. (2.1, 2.2)

3. $5k + 3j + j + 3i + k$ 3. _____

4. $c^2 + c + 2c$ 4. _____

5. $2(d + 4) + 3d$ 5. _____

6. $15ax + x^2 + 4ax$ 6. _____

In Exercises 7 and 8, evaluate the expression when $m = 4$ and $n = 2$. (2.1, 2.2)

7. $5(m + n) + 12$ 7. _____

8. $4m + 9(n + 1)$ 8. _____

In Exercises 9 and 10, solve the equation using mental math. (2.3)

9. $50 - q = 32$ 9. _____

10. $\dfrac{45}{r} = 9$ 10. _____

In Exercises 11–13, translate the verbal phrase into an algebraic expression. (2.4)

11. The sum of 5 times a number and 6   11. _____

12. The difference of 59 and a number   12. _____

13. The product of one number decreased by 6 and the same number increased by 6   13. _____

In Exercises 14–17, solve the equation. (2.4, 2.5)

14. $5.97 + v = 12$ 14. _____

15. $s - 145 = 292$ 15. _____

16. $1.5w = 7.5$ 16. _____

17. $\dfrac{x}{6} = 216$ 17. _____

In Exercises 18 and 19, write an equation and solve. (2.7)

18. You see a jacket in a discount store for $39.95. You see the same jacket in a sporting goods store for $9.49 more. How much does the jacket cost in the sporting goods store?

18. _____

19. The one-way distance between school and home is 0.73 miles. How many miles do you walk per 5-day week if you walk to and from school each day?

19. _____

20. Imagine that the amount of money in your wallet increased by $12.75 is $30.00. Write an algebraic equation for the verbal model. Then solve. (2.7)

20. _____

21. Write a verbal model and an algebraic equation that describes a real-life situation. Then solve the equation. (2.7)

21.

In Exercises 22–25, solve the inequality. (2.9)

22. $h + 27 \geq 46$

22. _____

23. $g - 54.7 < 24.5$

23. _____

24. $54 \leq 6x$

24. _____

25. $\dfrac{e}{4} > 6.4$

25. _____

For Exercises 26–28, imagine that you are making a 1387-mile road trip from Los Angeles, California to Dallas, Texas. The first day, you drive 375 miles. The second day, you drive 438 miles. (2.8)

26. Write a verbal model that relates the total distance, the distance traveled, and the distance remaining in the trip.

26. _____

27. Write an algebraic model that can be used to find the distance remaining.

27. _____

28. How many miles remain in the trip from Los Angeles to Dallas?

28. _____

For Exercises 29–32, use the region at the right. (2.6, 2.7)

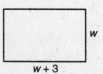

w

$w + 3$

29. Write an expression for the perimeter of the region.

29. _____

30. Find the perimeter of the region if $w = 12.5$.

30. _____

31. Write an expression for the area of the region.

31. _____

32. Find the area of the region if $w = 15$.

32. _____

3.2 Short Quiz

1. Plot the integers −3, −5, and 4 on the number line below.

1. _____

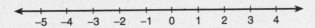

In Exercises 2 and 3, compare the integers using the symbols < or >.

2. −4 □ 3

2. _____

3. |−6| □ 5

3. _____

In Exercises 4 and 5, write the integer that represents the situation.

4. A loss of 12 yards

4. _____

5. A profit of $17

5. _____

6. Write and solve an equation to find the sum of 2 and −5.

6. _____

7. Write two sets of values for *r* and *s* that make the equation
   $r + s = -3$ true.

7.

In Exercises 8 and 9, imagine you are 2 stories underground in a parking garage. You ride an elevator up 5 floors.

8. Write an equation to describe the situation.

8. _____

9. Solve the equation. On what floor do you exit the elevator?

9. _____

3.4 Short Quiz

Name_____

Date _____

In Exercises 1 and 2, find the sum. Write your conclusion as an equation.

1. $-5 + 7 + (-3)$

1._____

2. $-9 + (-3) + 15$

2._____

In Exercises 3 and 4, simplify the expression. Then evaluate the expression when $r = 3$.

3. $6r + (-4) + 7$

3._____

4. $-5r + 8 + 2r$

4._____

In Exercises 5 and 6, find the difference. Write your conclusion as an equation.

5. $-3 - (-7)$

5._____

6. $-6 - 9$

6._____

In Exercises 7 and 8, simplify the expression.

7. $4s - 7s + 8$

7._____

8. $-6h - 5 - (-2h)$

8._____

9. Write a situation that could be solved using the equation $-3 + 8 = n$. Then solve.

9.

In Exercises 1 and 2, draw a number line and plot the numbers. (3.1)

1. $-3, 2, -5$

2. $6, -4, 0$

3. Write the opposite and absolute value of -10.

4. Choose four numbers between -5 and $+5$. Write them in order from least to greatest.

In Exercises 5 and 6, solve the equation. (3.2)

5. $-3 + u = 12$

6. $7 + v = -2$

In Exercises 7 and 8, evaluate the expression when $d = 6$ and $e = -8$. (3.1, 3.4)

7. $|d| - e$

8. $d - |e|$

9. Simplify the expression $8j + (-5j) - 6k$ when $j = 3$ and $k = 2$. (3.4)

In Exercises 10 and 11, use the bar graph showing the changes in Carrie's bank account for June through December. (3.1, 3.3)

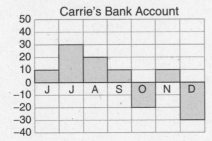

Carrie's Bank Account

10. During which two months did Carrie's balance change the most? Explain your reasoning.

11. Write an equation for the overall change in Carrie's account for June through December. Then solve.

1. _____

2. _____

3. _____

4. _____

5. _____

6. _____

7. _____

8. _____

9. _____

10. _____

11. _____

In Exercises 1 and 2, order the numbers from least to greatest. (3.1)

1. $5, -6, 3, |4|$

1._____

2. $-2, 7, 0, |-5|$

2._____

3. Write the opposite and absolute value of -8.

3._____

4. Choose four numbers between -4 and $+4$. Draw them on a number line.

4.

In Exercises 5 and 6, solve the equation. (3.2)

5. $4 + t = -7$

5._____

6. $-3 + g = -5$

6._____

In Exercises 7 and 8, evaluate the expression when $m = 5$ and $n = 7$. (3.1, 3.4)

7. $|m| - n$

7._____

8. $n - |m|$

8._____

9. Simplify the expression $12c + (-7c) - 2d$ when $c = 2$ and $d = 4$. (3.4)

9._____

In Exercises 10 and 11, use the bar graph showing the change in Jeremy's 50-meter freestyle swim times from his average time for one week. His average time is 35.00 (35 seconds). (3.1, 3.3)

Jeremy's 50-m Freestyle Times

10. On what day did Jeremy's time change the most? Explain your reasoning.

10.

11. On which day was Jeremy's time the fastest? What was it?

11._____

3.6 Short Quiz

Name _____

Date _____

In Exercises 1 and 2, find the product. Write your conclusion as an equation.

1. $6 \cdot (-4)$

1._____

2. $(-5)(-8)(2)$

2._____

3. Evaluate the expression sw^2 when $s = 5$ and $w = 2$.

3._____

In Exercises 4–7, use mental math to solve the equation.

4. $-5x = 35$

4._____

5. $-7m = -49$

5._____

6. $\dfrac{c}{3} = -6$

6._____

7. $\dfrac{72}{y} = -8$

7._____

In Exercises 8 and 9, evaluate the expression.

8. $\dfrac{90}{-5}$

8._____

9. $\dfrac{-459}{-9}$

9._____

10. Find the average of $-3, 4, -5, 6, -7,$ and 8.

10._____

11. List 6 numbers whose average would be -20.

11._____

3.8 Short Quiz

Name_____

Date _____

In Exercises 1–4, solve the equation.

1. $3x = -15$

1._____

2. $\dfrac{a}{-5} = -5$

2._____

3. $3.5 + z = -4.2$

3._____

4. $f - 4.9 = -16.7$

4._____

5. Write an algebraic equation for the sentence 15 is the sum of -10 and m. Then solve.

5._____

In Exercises 6–8, plot and label the points on a single coordinate plane. Identify the quadrant in which each point lies.

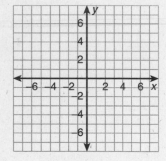

6. $I(5, -7)$

6._____

7. $N(-3, 4)$

7._____

8. $T(6, 1)$

8._____

9. Plot a point R in the quadrant that has no other points. In which quadrant is point R? What ordered pair identifies its location?

9._____

In Exercises 1 and 2, order the integers from least to greatest. (3.1)

1. $-5, 7, 3, 0, -8$ 1._____

2. $0, 8, -4, 3, -5$ 2._____

3. State the opposite and the absolute value of 63.   3._____

In Exercises 4 and 5, write the integer associated with the phrase. (3.1)

4. 30 feet below sea level   4._____

5. A $45 profit   5._____

In Exercises 6–11, evaluate the expression. (3.2–3.6)

6. $14 + (-6) + (-9)$ 6._____

7. $-43 + |-37|$ 7._____

8. $19 - (-46)$ 8._____

9. $21 \cdot (-3)$ 9._____

10. $-68 \div -4$ 10._____

11. $\dfrac{148}{-37}$ 11._____

In Exercises 12–14, simplify the expression. Then evaluate if $y = 3$ and $z = 2$. (3.3–3.5)

12. $15y + (-10y) + 4$ 12._____

13. $-4y + 8y - 3z$ 13._____

14. $-5z - (-15z) + 5$ 14._____

In Exercises 15 and 16, find the average. (3.6)

15. $-10, 5, 7, -9, 3, -8$

15._____

16. $-46, -10, 50, -14$

16._____

In Exercises 17 and 18, write an algebraic equation for the statement. Then solve the equation. (3.7)

17. The product of a number and -6 is -24.

17._____

18. The difference of a number and -5 is 9.

18._____

In Exercises 19–22, solve the equation.

19. $r + 8 = -7$

19._____

20. $s - 10 = -23$

20._____

21. $-8t = -56$

21._____

22. $\dfrac{u}{-6} = 9$

22._____

In Exercises 23–26, imagine that the temperature in Rockville at 7:00 AM is $-7°C$. By 12:00 noon, the temperature increases to 13°C, but it falls 3°C by 6:00 PM. (3.4, 3.6, 3.7)

23. How much does the temperature increase between 7:00 AM and 12:00 noon?

23._____

24. What is the average hourly temperature gain between 7:00 AM and 12:00 noon?

24._____

25. What is the average hourly temperature drop between 12:00 noon and 6:00 PM?

25._____

26. Decide how much the temperature might drop between 6:00 PM and midnight. Write an equation to show how you would find the 12:00 midnight temperature reading. Then solve.

26._____

In Exercises 27–28, consider the following: You are planning a rummage sale to raise money for a field trip. The expenses are $10 for flyers, $35 for advertising, and $50 for table rentals. You take in a total of $325.00. (3.7)

27. Use the verbal model profit = income − expenses to write an algebraic model for the profit. Solve to find the profit.

27._____

28. If the profit is split evenly among 25 students, how much does each student get towards the field trip?

28._____

In Exercises 29–32, plot each point on the coordinate graph and name its quadrant. (3.8)

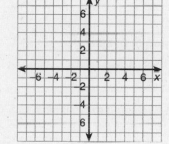

29. R: (5, 1)

29._____

30. S: (−5, −1)

30._____

31. T: (−5, 1)

31._____

32. U: (5, −1)

32._____

33. Connect $RSTU$. What type of figure is it?

33._____

34. What is the perimeter of $RSTU$?

34._____

35. What is the area of $RSTU$?

35._____

In Exercises 36–38, determine whether the ordered pair is a solution of the equation. (3.8)

36. $y = -3x$; (3, 1)

36._____

37. $x + 3 = y$; (−1, 2)

37._____

38. $y + x = -2$; (4, −6)

38._____

1. Which shows the relationship between 0, 1, and -3? (3.1)

   a. $0 < 1 < -3$ **b.** $-3 < 1 < 0$

   c. $-3 > 0 > 1$ **d.** $-3 < 0 < 1$

   1._____

2. Which shows the integers in order from least to greatest? (3.1)

   a. $0, -1, 2, -3, 4$ **b.** $-1, -3, 0, 2, 4$

   c. $-3, -1, 0, 2, 4$ **d.** $4, 2, 0, -1, -3$

   2._____

In Exercises 3 and 4, which phrase could be represented by the integer? (3.1)

3. $+7$

   a. 7 feet above sea level   **b.** 7 leagues under the sea

   c. 7 more than x **d.** x more than 7

   3._____

4. -10

   a. a profit of $10   **b.** x less than 10

   c. 10 less than x **d.** a loss of $10

   4._____

In Exercises 5 and 6, what is the value of the evaluated expression? (3.3, 3.4)

5. $42 + (-6) + (-36)$

   a. 84   **b.** 0   **c.** 12   **d.** 72

   5._____

6. $-84 - (-19)$

   a. -65 **b.** 103   **c.** -103 **d.** 65

   6._____

In Exercises 7 and 8, simplify the given expression. (3.4)

7. $4x - 3xy + (-5x)$

   a. $9x - 3xy$ **b.** $-3xy - 9x$

   c. $60x^3y$ **d.** $-3xy - x$

   7._____

8. $-6y + 4x + 7x - (-3y)$

   a. $-9y + 11x$ **b.** $3y + 11x$

   c. $11x - 3y$ **d.** $18y^2 + 28x^2$

   8._____

In Exercises 9 and 10, what is the value of the expression when $a = 3$ and $b = 2$? (3.3–3.5)

9. $4ab - 5a$ 9._____

   a. 39   **b.** 9   **c.** -9 **d.** 5

10. $-8b + (-3ab)$ 10._____

   a. -34 **b.** 2   **c.** -2 **d.** 34

In Exercises 11 and 12, what is the average of the numbers? (3.6)

11. $-4, 10, -3, 20, -1, 8$ 11._____

   a. -5 **b.** 5   **c.** -7.5 **d.** 7.5

12. $-42, -36, -52, -46, -42, -46$ 12._____

   a. -44 **b.** -42 **c.** 44   **d.** -46

In Exercises 13 and 14, which is the algebraic model for the statement? (3.7)

13. The difference of a number and -14 is 37.   13._____

   a. $n + (\ 14) = 37$ **b.** $n - (-14) = 37$

   c. $n \cdot (-14) = 37$ **d.** $\dfrac{n}{-14} = 37$

14. The product of -6 and a number is -58.   14._____

   a. $6n = -58$ **b.** $-6n = -58$

   c. $\dfrac{-6}{-58} = n$ **d.** $\dfrac{-6}{n} = -58$

In Exercises 15–17, solve the equation. (3.8)

15. $d - (-4) = 15$ 15._____

   a. $d = 19$ **b.** $d = -19$ **c.** $d = -11$ **d.** $d = 11$

16. $\dfrac{t}{-8} = -16$ 16._____

   a. $t = 128$ **b.** $t = 2$ **c.** $t = \dfrac{1}{2}$ **d.** $t = -24$

17. $-4r = 316$ 17._____

   a. $r = 1264$ **b.** $r = -1264$ **c.** $r = -79$ **d.** $r = 79$

In Exercises 18 and 19, consider the following: You keep track of the daily
high temperatures (in degrees Celsius) for one week during the winter. Your
records show daily high temperatures of 3°C, −6°C, −9°C, 0°C, −8°C, 1°C,
and 12°C. (3.4, 3.6)

18. What is the average high temperature for the week?   18._____

   a. 0°C   **b.** 5.4°C   **c.** −1°C   **d.** 1°C

19. What is the difference between the highest and lowest temperatures   19._____
for the week?

   a. 21°C   **b.** 3°C   **c.** −3°C   **d.** −21°C

In Exercises 20–23, in which quadrant would the point be located on a
coordinate graph? (3.8)

20. (−2, 4)   20._____

   a. Quadrant 1   **b.** Quadrant 2   **c.** Quadrant 3   **d.** Quadrant 4

21. (4, −6)   21._____

   a. Quadrant 1   **b.** Quadrant 2   **c.** Quadrant 3   **d.** Quadrant 4

22. (7, 4)   22._____

   a. Quadrant 1   **b.** Quadrant 2   **c.** Quadrant 3   **d.** Quadrant 4

23. (−5; −6)   23._____

   a. Quadrant 1   **b.** Quadrant 2   **c.** Quadrant 3   **d.** Quadrant 4

24. Which graph shows the points referred to in Exercises 20–23? (3.8)   24._____

a. 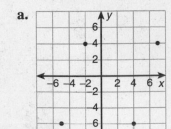   **b.** 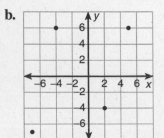   **c.** 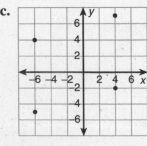   **d.**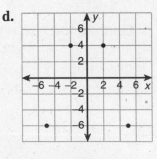

25. What is the shape of the figure created by connecting the points in   25._____
Exercise 24? (3.8)

   a. rectangle   **b.** square

   c. trapezoid   **d.** parallelogram

Form C
(Page 1 of 3 pages)

Name_____

Date _____

In Exercises 1 and 2, plot the numbers on a number line. (3.1)

1. $7, -3, 3, 1, -5$

2. $-4, -1, -5, 1, -2$

In Exercises 3 and 4, write a phrase associated with the integer. (3.1)

3. -12

4. $+12$

5. List five numbers that would have an average of 0.

In Exercises 6–10, evaluate the expression. (3.3–3.6)

6. $45 + (-57) - (-15)$

7. $-34 \cdot -9$

8. $(27)(-3)(-5)$

9. $\dfrac{385}{-15}$

10. $-4.29 \div -0.3$

In Exercises 11 and 12, write an algebraic equation for the statement. Then solve the equation. (3.7)

11. The difference of a number and 5 is -22.

12. The quotient of a number and -3 is -15.

1. _____

2. _____

3. _____

4. _____

5. _____

6. _____

7. _____

8. _____

9. _____

10. _____

11. _____

12. _____

In Exercises 13 and 14, simplify the expression. Then evaluate if $r = 2$ and $s = -5$. (3.3–3.5)

13. $-4rs + (-rs) - 2r^2$

13._____

14. $3rs + (-2rs) + 4r$

14._____

In Exercises 15–18, solve the equation. (3.7)

15. $y - (-15) + 19 = 27$

15._____

16. $-23 + g = 2$

16._____

17. $-0.6b = 12$

17._____

18. $\dfrac{t}{0.5} = -12.5$

18._____

In Exercises 19–22, consider the following: You are in the window-cleaning business. For each job, you spend $6.30 for supplies and $8.75 for the machine rental. You make a $27.90 profit on each job. (3.7)

19. Write an equation to show what you charge for a job. Then solve.

19._____

20. If it takes you 4.5 hours to complete a job, what is your hourly rate for labor (not including the charges for supplies and rent)?

20._____

21. Suppose you wanted to earn $7.00 per hour for labor. What would you need to charge per job to earn this rate based on a 4.5 hour job?

21._____

22. Based on an estimate of 4.5 hours per job, you chart the change in time spent on 6 new jobs. You record the change as an increase or decrease from your average time. What is your average time per job for these six jobs?

22._____

	Job 1	Job 2	Job 3	Job 4	Job 5	Job 6
Change	0	-0.5	$+0.5$	-0.25	-0.75	-0.5

For Exercises 23–28, plot four points on the graph so that one point is in each quadrant. Label the points A, B, C, and D. (3.8)

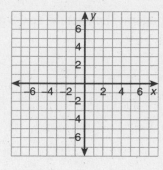

23. What are the coordinates of point A?

23._____

24. What are the coordinates of point B?

24._____

25. What are the coordinates of point C?

_____

26. What are the coordinates of point D?

26._____

27. Connect ABCD to form a polygon. What type of figure is it?

27._____

28. Draw a rectangle around ABCD. Find the perimeter and area of the rectangle.

28._____

29. Construct a table that lists six solutions of the equation $x + y = 0$. Then plot the points and describe the pattern. (3.8)

29.

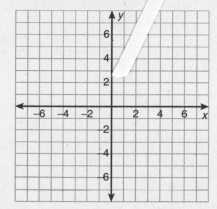

In Exercises 1–6, describe the pattern. Then list the next 3 numbers or letters. (1.1, 1.8)

1. 50, 45, 40, 35

2. $\frac{1}{2}, \frac{2}{3}, \frac{3}{4}, \frac{4}{5}$

3. A, 1, B, 2

4. Z, A, Y, B

5. 1,000,000; 100,000; 10,000; 1,000

6. 100, 90, 81, 73

1. _____

2. _____

3. _____

4. _____

5. _____

6. _____

In Exercises 7–12, evaluate the expression using a calculator. Round to two decimal places when necessary. (1.3)

7. 9^7

8. $\left(\frac{4}{5}\right)^6$

9. $\sqrt{55}$

10. $\sqrt{500}$

11. $(2.7)^5$

12. $\sqrt{43.27}$

7. _____

8. _____

9. _____

10. _____

11. _____

12. _____

In Exercises 13–15, evaluate the expression using a calculator. Round to two decimal places when necessary. (1.2–1.4)

13. $135 - 35 \div 7 \cdot 9$

13. _____

14. $45 \div (12 + 3) + 3^5$

14. _____

15. $(15 + 7) \cdot 21 - 18 - 4^4$

15. _____

In Exercises 16–24, evaluate the expression. (1.4, 3.1–3.6)

16. $|7| - |-4|$

16. _____

17. $-|-8| + |5|$

17. _____

18. $-4 + 12 - 8 + 7$

18. _____

19. $36 - 5 + (-13) + 50$

19. _____

20. $(-4)(13)$

20. _____

21. $\dfrac{-140}{-4}$

21. _____

22. $(-5)(-6)(-7)$

22. _____

23. $2^5 + (9 - 6)^2 - 5$

23. _____

24. $(3 - 7)^3 \div 8 - 12$

24. _____

In Exercises 25–29, decide whether the figure is a polygon. If it is, name it. If it is not, explain. (1.7)

25.

25. _____

26.

26. _____

27.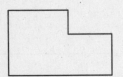

27. _____

28.

28. _____

29.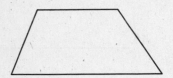

29. _____

30. Plot the numbers $-2, 4, 0$, and 2 on the number line below. (3.1)

30.

```
<———+———+———+———+———+———+———+———>
    -2  -1   0   1   2   3   4
```

In Exercises 31–33, order the numbers from least to greatest. (3.1)

31. $-3, 5, -8, 1$

31. _____

32. $|-2|, 0, |3|, -1$

32. _____

33. $3, |-2|, -4, |1|$

33. _____

In Exercises 34–37, use the Distributive Property to write an equivalent expression. Then evaluate it when $s = 5$, $a = 2$, and $m = 3$. (1.5, 2.1)

34. $4(s + a + 6)$

34._____

35. $7(s + a + m + 1)$

35._____

36. $3(m + 5a)$

36._____

37. $6(3a - m)$

37._____

In Exercises 38–43, rewrite and/or simplify the expression when possible. Then evaluate it when $r = 6$, $p = 3$, and $n = 4$. (1.5, 2.1, 2.2, 3.1–3.5)

38. $3r + 2p + 4n$

38._____

39. $8r - 2n - 4r - 5$

39._____

40. $4(2r + 3p - n) + p$

40._____

41. $8(2r + 3r) - 20$

41._____

42. $-12p + 3n - (-5p) - n$

42._____

43. $n(3r + 2) + |-p|$

43._____

In Exercises 44–46, plot and label the ordered pairs on the coordinate plane at the right. Then identify the quadrant for each plotted point. (3.8)

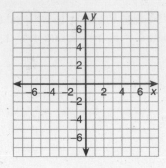

44. $A(4, -5)$ 44._____

45. $B(-3, -2)$ 45._____

46. $C(5, 1)$ 46._____

47. Plot a point D in the quadrant that has no point. Identify the ordered   47._____
    pair and the quadrant in which the point lies.

In Exercises 48–55, solve the equation. (2.3–2.5, 3.7)

48. $a + 9 = 23$ 48._____

49. $z + (-12) = 20$ 49._____

50. $b - 5 = -14$ 50._____

51. $y - (-9) = 16$ 51._____

52. $3c = 72$ 52._____

53. $-56 = 4d$ 53._____

54. $\dfrac{e}{-8} = -8$ 54._____

55. $t \div (-17) = 3$ 55._____

In Exercises 56–60, solve the inequality. (2.9)

56. $15 > f + 4$

57. $g - 12 > 5$

58. $n - 15 \leq 1$

59. $12h \leq 84$

60. $18 \leq i \div 3$

56._____

57._____

58._____

59._____

60._____

In Exercises 61–63, consider a square room that has an area of 256 square feet. (1.4)

61. What is the length of each side of the room?

62. The room has a tile floor. Each side has 24 square tiles. What is the length (in inches) of each tile?

63. What is the area of each tile?

61._____

62._____

63._____

In Exercises 64–66, you buy fine line markers for $0.49 each and wide line markers for $0.35 each. You purchase x markers of each type. (2.7, 2.8)

64. Write an expression for the total amount of money spent.

65. If you purchased 3 of each type of marker, how much would you spend?

66. How much more would it cost to purchase 10 fine line markers than 10 wide line markers?

64._____

65._____

66._____

In Exercises 67–73, use the diagram below. (1.4, 3.8)

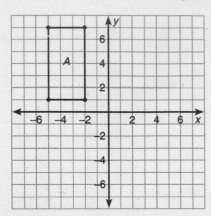

67. Name the coordinates of the vertices of figure *A*.   67. _____

68. Name the polygon that describes figure *A*.   68. _____

69. Name the quadrant in which figure *A* lies.   69. _____

70. Find the perimeter and area of figure *A*.   70. _____

71. Draw a polygon that lies entirely in Quadrant 1. Make one vertex at (7, 7).   71. _____

72. Name the coordinates of the other vertices of the polygon.   72. _____

73. Name the polygon.   73. _____

In Exercises 74–77, write an algebraic equation or inequality for the sentence. Then solve the equation or inequality. (2.6, 2.9)

74. -3 is the sum of a number and 15.

74._____

75. The difference of a number and 8 is less than or equal to 24.

75._____

76. The product of a number and 9 is greater than 144.

76._____

77. The quotient of a number and 7 is 8.

77._____

For Exercises 78–80, imagine that you have opened a savings account. You saved $15 in October and $43 in November. You spent $49 in December, then saved $12 in January. (3.3, 3.7)

78. Write an algebraic equation to determine the current balance in your account.

78._____

79. What is your account balance?

79._____

80. If you save $13 in February and $10 in March, will you have enough money to buy a school yearbook for $40?

80._____

81. You purchase 14 picture frames at a discount store. You buy s 5-by-7-inch frames and l 8-by-10-inch frames. List all the possible values of s and l in a table. (3.8)

81.

4.2 Short Quiz

Name_____

Date _____

In Exercises 1–7, solve the equation.

1. $4x + 13 = 41$

1._____

2. $-3y - 5 = 13$

2._____

3. $\dfrac{z}{8} + 4 = 3$

3._____

4. $9a - 150 = 201$

4._____

5. $3y + 6y - 7 = 11$

5._____

6. $7(w + 5) = 84$

6._____

7. Describe a real life situation that can be modeled by the equation $3x + 5 = 80$. Then solve the equation.

7.

For Exercises 8 and 9, imagine that you win *x* tickets each time you play skeeball. You play 6 games of skeeball on Saturday and 8 games of skeeball on Sunday. After giving two tickets away, you find that you have 68 tickets.

8. Write an equation that represents the number of tickets you have.

8._____

9. Solve the equation. How many tickets do you win each time you play skeeball?

9._____

4.4 Short Quiz

Name_____

Date _____

In Exercises 1–7, solve the equation.

1. $3b - 15 = 6$

1._____

2. $-\dfrac{1}{4}c + 5 = 1$

2._____

3. $\dfrac{3}{5}d - 2 = 4$

3._____

4. $-6(f + 3) = 42$

4._____

5. $g + 5(g - 4) = 4$

5._____

6. $7(h - 1) + 5h = 5$

6._____

7. $-6(j + 3) - 4j = 2$

7._____

In Exercises 8–10, consider a rectangle with a width of w and a length of $w + 3$.

8. Make a sketch of the rectangle.

8._____

9. Write an algebraic model for finding the perimeter of the rectangle.

9._____

10. Choose a whole number greater than or equal to 10 for your perimeter. Then solve to find the width and length of the rectangle using the perimeter you chose.

10._____

1. A rectangle has a perimeter of 32 inches. Its width, w, is 4 inches shorter than its length. Sketch the rectangle and find its dimensions. (4.3)

1._____

2. Find the reciprocal of $-\frac{4}{3}$. 3)

2._____

3. Write any negative number. Then write its reciprocal. (4.3)

3._____

In Exercises 4–8, solve the equation. (4.4)

4. $3e - 7 = 11$

4._____

5. $5y + 3 = 18$

5._____

6. $-\frac{1}{2}b + 9 = 5$

6._____

7. $10p - p + 2 = -7$

7._____

8. $35 = 5(n + 2)$

8._____

In Exercises 9–11, write an equation that represents the verbal sentence or problem. Then solve the equation. (4.1)

9. The sum of $4x$ and 7 is 27.

9._____

10. 22 is the difference between $3f$ and 50.

10._____

11. Boswell's Burger Barn sold 495 hamburgers today. The number sold with cheese was 6 less than twice the number sold without cheese. How many hamburgers with cheese and how many without cheese were sold? (4.3)

11._____

1. A rectangle has a perimeter of 48 feet. Its length, l, is 6 feet shorter than its width. Sketch the rectangle and find its dimensions. (4.3)

1. _____

2. Find the reciprocal of $-\frac{1}{8}$. (4.3)

2. _____

3. Write any positive number. Then write its reciprocal. (4.3)

3. _____

In Exercises 4–8, solve the equation. (4.1–4.4)

4. $5n - 9 = 71$

4. _____

5. $2x + 1 = 11$

5. _____

6. $\frac{1}{2}r + 3 = 9$

6. _____

7. $4(y + 2) = 24$

7. _____

8. $-8y - 11 = 13$

8. _____

In Exercises 9–11, write an equation that represents the verbal sentence or problem. Then solve the equation. (4.1)

9. -17 is the sum of 10 and $9h$.

9. _____

10. The difference of $3x$ and 8 is 25.

10. _____

11. A new oil tank holds 35 barrels of oil more than an old tank. Together they hold 365 barrels of oil. How much does each tank hold? (4.3)

11. _____

4.6 Short Quiz

Name_____

Date _____

In Exercises 1–3, solve the equation.

1. $3x + 16 = 7x$

2. $-4t - 2 = 2t + 4$

3. $6(3 + w) = 4 - w$

1._____

2._____

3._____

In Exercises 4 and 5, use the model below.

4. Write an equation implied by the linear model. Then solve it.

4._____

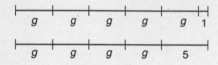

5. Draw a rectangle with an equivalent perimeter.

5.

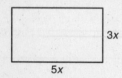

In Exercises 6–8, use the following: Tim has $12 in the bank and adds $3 per week. His sister, Dominique, has $60 in the bank and spends $5 per week. In how many weeks will Tim and Dominique have the same amount in the bank?

6. Make a table to show how to solve the problem.

6.

7. Write an algebraic model to solve the problem.

7._____

8. Solve the algebraic equation. In how many weeks will Tim and Dominique have the same amount in the bank?

8._____

4.8 Short Quiz

Name_____

Date _____

In Exercises 1–4, use a calculator to solve the equation. Round your result to two decimal places.

1. $4x + 15 = 5$ 1._____

2. $32f - 15 = -27$ 2._____

3. $0.3(3.5t + 4.2) = 12.75$ 3._____

4. $5.82x - 4 = 1.9x + 3.06$ 4._____

In Exercises 5 and 6, use the following figure.

5. This regular pentagon has a perimeter of 88 feet. What is the length of each side?   5._____

6. Choose another perimeter for the figure. Then find the length of each side using the perimeter you chose.   6._____

In Exercises 7 and 8, use the figure at the right.
Assume that the sum of the angles equals 180°.

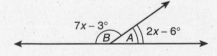

7. What is the value of x?   7._____

8. What is the measure of angle A? angle B?   8._____

In Exercises 1 and 2, state the inverse. (4.1)

1. Dividing a number by 7.

1. _____

2. Adding -5.2 to a number.

2. _____

In Exercises 3–5, solve the equation. (4.1, 4.3)

3. $3x + 9 = 27$

3. _____

4. $-5y - 4 = 76$

4. _____

5. $10 - \frac{1}{4}b = 9$

5. _____

In Exercises 6–8, write the reciprocal of the number. (4.3)

6. $\frac{5}{6}$

6. _____

7. -4

7. _____

8. $\frac{1}{10}$

8. _____

In Exercises 9–12, solve the equation. (4.2, 4.3, 4.5)

9. $4c + 7 - 6c = 23$

9. _____

10. $13d - 53 = 9d + 19$

10. _____

11. $\frac{5}{6}e + 18 - \frac{1}{6}e = 6$

11. _____

12. $25 - \frac{3}{7}f = 19 + \frac{4}{7}f$

12. _____

In Exercises 13 and 14, use the diagram below. (4.1, 4.5)

13. Write the equation implied by the diagram.

13._____

14. Solve to find the value of r.

14._____

In Exercises 15–19, solve the equation. (4.3–4.5)

15. $5(3m - 4) = -15$

15._____

16. $\frac{1}{2}(h - 6) = 19$

16._____

17. $\frac{1}{3}(5n + 12) + \frac{1}{3}n = -1$

17._____

18. $3(y + 4) - 10 = 66 - 5y$

18._____

19. $4(g - 2) = 2(g + 8)$

19._____

In Exercises 20 and 21, the sum of the two angles shown is 90°. (4.8)

20. What is the value of x?

20._____

21. What is the measure of each angle?

21._____

22. Write an equation with a variable so that when it is solved, the value of the variable is 5. Then show the solution steps. (4.1–4.5)

22._____

In Exercises 23 and 24, use the figures. (4.5)

23. Find the value of v so that the triangle and the square have the same perimeter. What is the value of v?

23._____

24. What is the perimeter of both figures?

24._____

In Exercises 25–27, use a calculator to solve the equation. Round your result to 2 decimal places. (4.7)

25. $5.6h - 6.8 = 43.27$

25._____

26. $1.4(u - 3.7) = 17.08$

26._____

27. $3.05(2w + 4.3) = -20.07$

27._____

In Exercises 28–30, you and your friend are swimming laps. Your friend jumps right in and swims 4 laps every five minutes. Ten minutes later, you begin swimming 6 laps every 5 minutes. (4.6)

28. After how many minutes will you catch up to your friend?

28._____

29. When you catch up, how many laps will you have swum?

29._____

30. Complete the table.

30.

Minutes	5	10	15	20	25	30
Your laps	0					
Friend's laps	4					

For Exercises 1 and 2, what is the inverse? (4.1)

1. Subtracting 15.4 from a number

   a. adding 15.4 to a number
   b. multiplying 15.4 to a number
   c. dividing a number by 15.4
   d. adding two numbers for a sum of 15.4

1._____

2. Multiplying a number by -3.2

   a. multiplying a number by 3.2
   b. adding a number to -3.2
   c. dividing a number by -3.2
   d. subtracting a number from -3.2

2._____

In Exercises 3–5, what is the solution of the equation? (4.1, 4.3)

3. $15 = 2p \quad 3$

   a. $p = 6$ b. $p = -9$ c. $p = 9$ d. $p = -6$

3._____

4. $54 - 2x = 20$

   a. $x = 37$ b. $x = 17$ c. $x = -17$ d. $x = -37$

4._____

5. $\frac{1}{4}j - 12 = 13$

   a. $j = 4$ b. 25   c. -4 d. 100

5._____

In Exercises 6–8, what is the reciprocal? (4.3)

6. $-\dfrac{9}{7}$

   a. $\dfrac{9}{7}$ b. $\dfrac{7}{9}$ c. $-\dfrac{12}{7}$ d. $-\dfrac{7}{9}$

6._____

7. $\dfrac{1}{5}$

   a. 5   b. $-\dfrac{1}{5}$ c. -5 d. $1\dfrac{1}{5}$

7._____

8. -23

   a. 23   b. $-\dfrac{1}{23}$ c. $\dfrac{1}{23}$ d. $-\dfrac{2}{3}$

8._____

In Exercises 9–11, what is the solution of the equation? (4.2, 4.3, 4.5)

9. $5x - 14 + 2x = 70$

   a. $x = -12$ **b.** $x = 12$ **c.** $x = 28$ **d.** $x = 8$

   9._____

10. $9r + 7 = 4r - 8$

   a. $r = -\frac{13}{15}$ **b.** $r = 5$ **c.** $r = -\frac{1}{5}$ **d.** $r = -3$

   10._____

11. $\frac{5}{8}n + 12 = \frac{3}{8}n + 4$

   a. -32 **b.** 32 **c.** $n = -8$ **d.** -16

   11._____

In Exercises 12 and 13, use the figure shown at the right. The sum of the angles is 180°. (4.8)

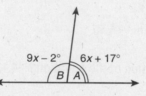

$9x - 2°$ $6x + 17°$

B A

12. What is the value of x?

   a. $x = 10$ **b.** $x = 12$ **c.** $x = 15$ **d.** $x = 11$

   12._____

13. What are the measures of the two angles?

   a. 90° and 90°   **b.** 116° and 64°

   c. 83° and 97°   **d.** 96° and 84°

   13._____

In Exercises 14–17, solve the equation. (4.3–4.5)

14. $2b + 3(b - 7) = 44$

   a. $b = 13$ **b.** $b = -13$ **c.** $b = \frac{44}{5}$ **d.** $b = 10$

   14._____

15. $4(2y + 9) + 7y = -24$

   a. $y = -\frac{33}{9}$ **b.** -12 **c.** -6 **d.** -4

   15._____

16. $\frac{1}{4}(n - 16) = 2$

   a. $n = 24$ **b.** $n = 8$ **c.** $n = 12$ **d.** $n = 72$

   16._____

17. $5(b - 3) = 7(b - 2)$

   a. $\frac{1}{2}$ **b.** 2 **c.** -2 **d.** $-\frac{1}{2}$

   17._____

For Exercises 18 and 19, use the figures below. Assume all sides of each figure are congruent. (4.5)

18. For what value of h will the figures have equal perimeters?

   a. $h = \frac{12}{8}$ **b.** $h = 6$ **c.** $h = 4$ **d.** $h = \frac{2}{3}$

18._____

19. What is the perimeter of each figure?

   a. 6   **b.** 20   **c.** 30   **d.** 3

19._____

For Exercises 20–22, use a calculator to solve the equation. Round your result to 2 decimal places. (4.7)

20. $-4.9u + 5.87 = 23.81$

   a. $u = 3.66$ **b.** $u = -3.66$ **c.** $u = 6.06$ **d.** -6.06

20._____

21. $5.32(b - 3) = 4.85$

   a. $b = -1.48$ **b.** $b = -3.91$ **c.** $b = 1.48$ **d.** $b = 3.91$

21._____

22. $8.1(z + 2.3) = 4.9(z - 4.2)$

   a. -12.26 **b.** -12.25 **c.** 12.25 **d.** 12.53

22._____

In Exercises 23 and 24, Pat earns $5 an hour as a waiter and Elan earns $3 an hour plus tips. They earned the same amount of money in the same number of hours when Elan earned $18 in tips. (4.6)

23. Which equation is the algebraic model for solving the problem?

   a. $5x + 3x = 18$ **b.** $18 - 3x = 5x$

   c. $3x + 18 = 5x$ **d.** $5x + 18 = 3x$

23._____

24. How many hours did Pat and Elan work?

   a. 4 hours   **b.** 2 hours   **c.** 8 hours   **d.** 9 hours

24._____

Name_____

Date_____

In Exercises 1 and 2, state the inverse.   (4.1)

1. Dividing a number by 3 and subtracting 7 from its quotient.

   1.

2. Adding 5 to a number and multiplying its sum by 6.

   2.

In Exercises 3–5, write the sentence as an equation. Then solve it.   (4.1, 4.3, 4.5)

3. 4 times a number increased by 13 is 69.

   3. _____

4. The sum of 32 and 7 times a number is -3.

   4. _____

5. One third a number decreased by 12 is 5.

   5. _____

6. Write an equation in its verbal form (as in Exercises 3–5). Then write its algebraic model and solve.

   6.

In Exercises 7 and 8, use the figure at the right. Assume that the sum of angles in any quadrilateral is 360°.   (4.8)

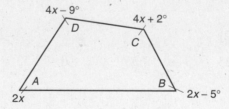

7. What is the value of x?

   7. _____

8. What is the measure of each angle in the figure?

   8. _____

In Exercises 9 and 10, state the reciprocal.   (4.3)

9. $-\dfrac{1}{5}$

   9. _____

10. 32

    10. _____

In Exercises 11–15, solve the equation. (4.3–4.5)

11. $3s + 7s = -5s + 9$

11._____

12. $\frac{5}{8}x + 12 = \frac{3}{8}x + 4$

12._____

13. $\frac{2}{5}n - 3 = \frac{9}{5}n + 9$

13._____

14. $3(a + 4) = 5(a - 6)$

14._____

15. $\frac{1}{2}(y - 16) = \frac{3}{4}y$

15._____

In Exercises 16 and 17, Suad Al-Kawas bought some blank cassettes for $3.50 each and half as many single cassettes for $2.50 each. (4.6)

16. Write an algebraic model that represents Suad's purchases if her total cost was $28.50. Then solve.

16._____

17. How much more did Suad spend on blank cassettes than on singles?

17._____

18. Find the value of x so that the perimeter of the square is 15 less than the perimeter of the equilateral triangle.

18._____

19. Your partner has difficulty solving the equation $7n - 2n + 3 = 5n + 3$. What would you say? What is the solution? (4.2, 4.5)

19.

In Exercises 20–23, you and your brother are busy boxing homemade cookies. You begin to box cookies at the rate of 2 boxes every 15 minutes. Your brother starts one-half hour later but can work at the rate of 1 box every 5 minutes. How many boxes of cookies will you each complete when your accomplishments are equal? **(4.6)**

20. Write an algebraic model that could be used to solve the problem. Then solve it.

20._____

21. What does the variable represent in your equation?

21._____

22. Complete the table.

22.

Time (hrs)	$\frac{1}{2}$	$\frac{3}{4}$	1	$1\frac{1}{4}$	$1\frac{1}{2}$	$1\frac{3}{4}$
You						
Your brother						

23. After how much time will your brother announce that he's boxed more cookies than you?

23._____

In Exercises 24 and 25, use a calculator to solve the equation. Round your result to 2 decimal places. **(...)**

24. $8.97(0.1x - 7.92 + 3.4x) = 10 - 3.25x$

24._____

25. $4.35e - 5.34e + 1.1 = 0.56e$

25._____

In Exercises 26–28, you are having notices professionally printed. Design costs are $19.00. The first hundred notices cost $5.50 and additional notices are sold in hundreds for $4.95. How many notices can you have printed if you have $50 to spend? **(4.6)**

26. Create a verbal model for the problem.

26.

27. Solve the problem using a table or an algebraic model.

27._____

28. Describe and explain your solution strategy. Why did you choose the method you did?

28.

5.2 Short Quiz

Name_____

Date _____

In Exercises 1–5, the picture graph shows a computer software store's sales record for one month.

November Sales

Education	🖫 🖫 🖫 🖫
Games	🖫 🖫 🖫 🖫 🖫
Self-Help	🖫 🖫 🖫
Business	🖫 🖫 🖫 🖫 🖫

🖫 = $500

1. How much money does one piece of software represent?

1._____

2. Estimate the amount of money spent in software games in November.

2._____

3. How much more money was spent on education software than on self-help software?

3._____

4. If one disk represented $250, how would the picture change?

4._____

5. Write a problem that can be solved by interpreting the graph. Then solve.

5._____

In Exercises 6–9, the histogram shows the scores earned by a group of students on a science test.

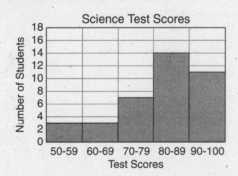

Science Test Scores

6. What interval contains the greatest number of students?

6._____

7. What intervals contain the same number of students?

7._____

8. How many scores were recorded on the histogram?

8._____

9. What conclusions can you make from the histogram?

9._____

Name_____

Date _____

In Exercises 1–5, use the line graph showing the prices of three stocks over a period of 8 weeks.

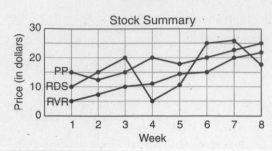

1. What are the units of the horizontal and vertical axes?

1.

2. What do the three lines represent in the graph?

2.

3. Which stock increased every week?

3._____

4. During which time periods did stock PP decrease?

4._____

5. Which stock would you purchase on week 9, based on the data in the line graph? Explain your answer.

5.

In Exercises 6–8, use the data below.

Hours of TV Watched in One Week (Middle School Students)

Time (to the nearest hour)	0–4	5–9	10–14	15–19	16–20	20–24
Students	5	8	9	6	8	2

6. What type of graph would best represent the data? Explain your answer.

6.

7. Construct a graph to show the data.

7.

8. What can you conclude from your graph?

8.

| **Form A**
(Use after Lesson 5.4)

Name_____

Date_____

In Exercises 1–3, use the graph at the right, which shows the attendance for the first six games of the Wildcats' Football team. (5.3)

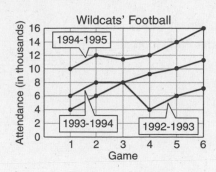

1. Which season had the best attendance record?

1._____

2. Which season had no decrease in attendance?

2._____

3. Which season probably had the best record? Explain your reasoning.

3.

4. Draw a graph that best represents tee shirt sales for the Patriots for the years given. Explain why you chose that type of graph. (5.4)

4.

1992: $325;   1993: $410;   1994: $475;   1995: $425;   1996: $500

In Exercises 5 and 6, use the data in the graph at the right, which shows the soft drink preferences of middle school students. (5.2)

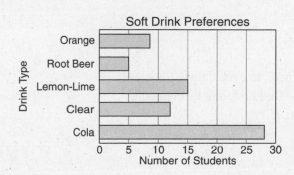

5. Which soft drink is most commonly favored?

5._____

6. If 8 more students had chosen lemon-lime, how would its total compare to the cola choice? Explain.

6.

In Exercises 1–3, use the graph, which shows concession sales for the "Powder Puff" football game. (5.3)

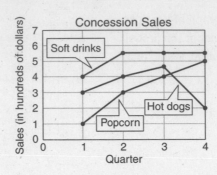

1. Which item had the steadiest increase in sales throughout the game?

1._____

2. Which item had the most consistent sales?

2._____

3. Choose one of the items and offer an explanation for its sales pattern throughout the game.

3.

4. Draw a graph that best represents ticket sales for the "Powder Puff" football game for the years given. Explain why you chose that type of graph. (5.4)

   1992: $575;   1993: $610;   1994: $800;   1995: $725;   1996: $850

4.

In Exercises 5 and 6, use the data in the graph, which shows the fruit preferences of middle school students. (5.2)

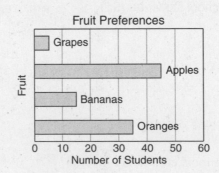

5. Which fruit is most commonly favored?

5._____

6. If 25 more students had chosen bananas, how would its total compare to the apples choice? Explain.

6.

Name _____

Date _____

In Exercises 1–4, use the bar graph at the right showing the weekly earnings for particular items in the school store.

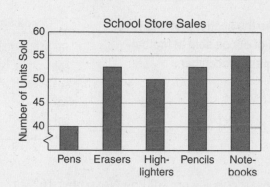

School Store Sales

1. Judging only from the length of the bars, compare the sales of pens and pencils.

1. _____

2. Use the scale to determine the answer to Exercise 1.

2. _____

3. Use the information in the graph to create another bar graph that is not misleading.

3. _____

4. If a pen salesman wanted to sell your school pens, which graph would he prefer to use? How about a pencil salesman? Explain.

4. _____

In Exercises 5 and 6, use the data showing the preferences of 30 students for a school mascot.

tiger	lion	dolphin	tiger	dolphin	eagle
shark	eagle	tiger	lion	eagle	tiger
eagle	shark	tiger	tiger	eagle	dolphin
eagle	eagle	lion	dolphin	eagle	shark
shark	lion	eagle	eagle	dolphin	tiger

5. Organize the data into a line plot.

5. _____

6. Which two animals received less votes combined than the number of votes received by the eagle?

6. _____

5.8 Short Quiz

Name_____

Date _____

In Exercises 1–4, use the scatter plot showing the number of cups of cocoa sold at football games in November and December.

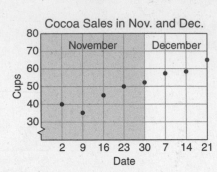

1. How many cups of cocoa were sold on November 16?

   1._____

2. How are the dates and sales related? Explain.

   2.

3. Draw a line on the scatter plot that appears to best fit the points.

   3._____

4. Use the result of Exercise 3 to estimate the cups of cocoa sold on December 28.

   4._____

In Exercises 5–8, find the probability of the spinner landing on the number.

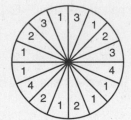

5. 1

   5._____

6. 2

   6._____

7. 3

   7._____

8. 4

   8._____

9. How many more sections would need to be labeled "1" in order for the probability of the spinner landing on it to be $\frac{1}{2}$?

   9._____

In Exercises 10–12, use the following statements to color the spinner.

10. Probability of red is $\frac{2}{8}$.

    10.

11. Probability of blue is $\frac{5}{8}$.

    11.

12. Probability of yellow is $\frac{1}{8}$.

    12.

In Exercises 1 and 2, use the picture graph showing the speeds of animals. (5.1)

(Source: 1997 World Almanac)

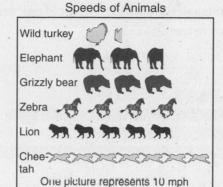

Speeds of Animals

Wild turkey
Elephant
Grizzly bear
Zebra
Lion
Chee-tah

One picture represents 10 mph

1. Estimate the speed of each animal.

1.

2. About how much faster is a cheetah than an elephant?

2._____

In Exercises 3 and 4, use the time line showing highlights from the first 300 years of American history. (5.1) *(Source: Rise of the American Nation)*

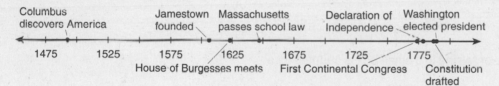

Columbus discovers America   Jamestown founded   Massachusetts passes school law   Declaration of Independence   Washington elected president

1475   1525   1575   1625   1675   1725   1775

House of Burgesses meets   First Continental Congress   Constitution drafted

3. Estimate the year that the Constitution was drafted.

3._____

4. Name two historical events in the eighteenth century.

4.

5. Organize the data below and represent your results graphically. Then explain why you used the type of graph you chose.

5.

Heights of Eighth Grade Students at Cabin John Middle School (in inches)

56	60	57	58	64	71	66	61
58	63	65	71	70	65	62	58
59	62	64	65	68	59	61	67

In Exercises 6 and 7, use the bar graph showing the number of swimmers who swim for the Stonebridge swim team. (5.2, 5.5)

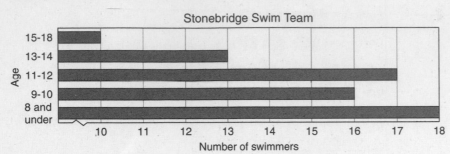

6. Without looking at the scale, compare the number of swimmers who are 8 and under with those who are in the 15–18 age group. Then read the scale and compare the actual numbers.

6.

7. Redraw the graph so that it is not misleading.

7.

In Exercises 8–11, the number of raisins in a small box is counted by a quality control group. (5.6)

27 28 35 34 29 28 27 29 33 35 34 29 30 31 27 29 28 29 30 33

8. Organize the data in a line plot.

8.

9. What is the number of raisins found in the greatest number of boxes?

9._____

10. How many boxes contained more than 30 raisins?

10._____

11. If you bought a small box of raisins, would you expect to find more or less than 30 raisins inside? Explain.

11._____

In Exercises 12–14, use the data below showing the age of a tree (in years) and its diameter (in inches). (5.7) *(Source: HBJ Science, 1987)*

(5, 1.0)   (8, 2.1)   (10, 2.7)   (16, 4.3)   (23, 4.8)   (30, 6.0)   (35, 7.2)   (40, 7.6)

12. Draw a scatter plot that relates the age of the tree with its diameter.

12.

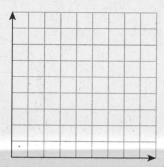

13. Does the scatter plot show a positive correlation, a negative correlation, or no correlation?

13._____

14. Based on the scatter plot, about how large would you expect the diameter of a 50-year-old tree to be?

14._____

In Exercises 15–18, a bag contains 3 red cubes, 1 white cube, and 8 blue cubes. You choose one cube from the bag without looking at the color. (5.8)

15. What is the probability that the cube will be red?

15._____

16. What is the probability that the cube will be blue?

16._____

17. What is the probability that the cube will be white?

17._____

18. What is the probability that the cube will be red or white?

18._____

In Exercises 19 and 21, you roll a regular number cube (die). (5.8)

19. What is the probability that you will roll a number less than 5?

19._____

20. For what numerals shown on the cube would the probability be $\frac{2}{6}$?

20._____

21. What is the probability that you roll an even number?

21._____

In Exercises 1–3, use the picture graph at the right showing the shoe preferences of middle school students. (5.1)

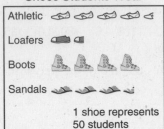

Shoes Students Wear

Athletic

Loafers

Boots

Sandals

1 shoe represents 50 students

1. How many students does each shoe represent?

   a. 25   **b.** 50   **c.** 10   **d.** 100

   1._____

2. How many students selected athletic shoes as their choice?

   a. 45   **b.** 450   **c.** 125   **d.** 225

   2._____

3. About how many more students wear boots than loafers?

   a. 15   **b.** 50   **c.** 125   **d.** 150

   3._____

In Exercises 4–6, use the bar graph at the right showing the automobiles sold in January. (5.2, 5.5)

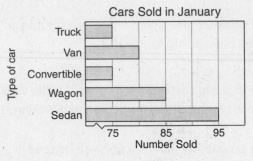

Cars Sold in January

Type of car

Truck
Van
Convertible
Wagon
Sedan

75   85   95

Number Sold

4. Without looking at the scale, how did the sales of sedans compare with truck sales?

   a. 5 times as many sedans are sold than trucks

   b. 5 times as many trucks are sold than sedans

   c. 5 more sedans are sold than trucks

   d. 5 more trucks are sold than sedans

   4._____

5. What is the actual difference between the sales of sedans and trucks?

   a. 20 more sedans than trucks   **b.** 40 more trucks than sedans

   c. 4 more sedans than trucks   **d.** 5 more sedans than trucks

   5._____

6. Why is the graph misleading?

   a. the scale is broken

   b. the horizontal and vertical axes are not equal

   c. the scale is too large

   d. the bars are the wrong length

   6._____

In Exercises 7–10, use the time line showing the life of George Washington. (5.1)

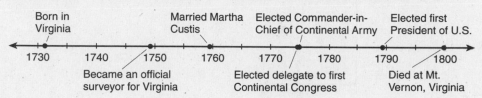

7. What is the time increment used in the time line?   7._____

   a. 5 years   **b.** 10 years   **c.** 1 year   **d.** 70 years

8. Approximately when did George Washington marry Martha Custis?   8._____

   a. 1754   **b.** 1758   **c.** 1760   **d.** 1748

9. Which event did not occur after 1750?   9._____

   a. Washington was elected delegate to first Continental Congress

   b. Washington was elected Commander-in-Chief of Continental Army

   c. Washington became an official surveyor for Virginia

   d. Washington died at Mt. Vernon, Virginia

10. About how old was George Washington when he died?   10._____

   a. about 63   **b.** about 67   **c.** about 73   **d.** about 77

In Exercises 11–13, use the line plot showing the number of letters in an eighth grade student's last name. (5.6)

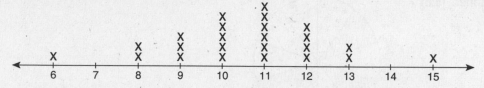

11. What is the number of letters found in the greatest number of names?   11._____

   a. 12   **b.** 10   **c.** 8   **d.** 11

12. How many names contain 10 letters or less?   12._____

   a. 11   **b.** 6   **c.** 4   **d.** 13

13. If you picked a name, about how many letters would you expect the   13._____
    name to contain?

   a. between 12 and 15   **b.** between 9 and 12

   c. between 6 and 9   **d.** about 12

14. What kind of correlation, if any, does the scatter plot show? (5.7)

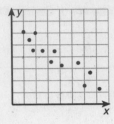

   a. positive correlation

   b. negative correlation

   c. no correlation

   d. correlation of 1

14. _____

In Exercises 15–17, use the graph showing the average temperature in Houston, Texas between October and April. (5.3)

(Source: 1997 World Almanac)

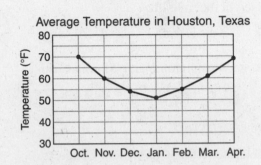

Average Temperature in Houston, Texas

15. What type of graph is this?

   a. bar   **b.** line plot   **c.** line   **d.** scatter plot

15. _____

16. In which month is the temperature lowest?

   a. October   **b.** November   **c.** January   **d.** April

16. _____

17. Which month has about the same average temperature as November?

   a. December   **b.** March   **c.** February   **d.** April

17. _____

In Exercises 18–20, use the spinner. (5.8)

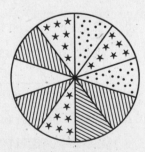

18. What is the probability of spinning a starred section?

   a. $\frac{1}{10}$ **b.** $\frac{3}{10}$ **c.** $\frac{3}{7}$ **d.** $\frac{7}{3}$

18. _____

19. What is the probability of spinning an unmarked section?

   a. $\frac{1}{10}$ **b.** $\frac{3}{10}$ **c.** $\frac{1}{9}$ **d.** $\frac{9}{10}$

19. _____

20. Which section has a probability of $\frac{2}{10}$ that it will be landed upon when spun?

   a. stars   **b.** stripes   **c.** plain   **d.** dots

20. _____

Form C
(Page 1 of 3 pages)

Name_____

Date _____

In Exercises 1–3, use the picture graph showing how eighth grade students from Lincoln Middle School get to school.

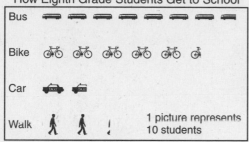

How Eighth Grade Students Get to School

1 picture represents 10 students

1. How many students does each picture represent?

1._____

2. Estimate the number of students who use each type of transportation shown on the graph.

2.

3. About how many more students ride the bus than ride a bike?

3._____

In Exercises 4–8, use the time line showing Important dates in California's history. (5.1)

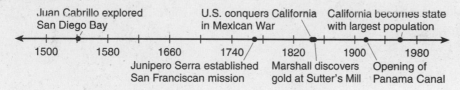

4. What is the increment used in the time line?

4._____

5. Estimate the year that Juan Cabrillo explored the San Diego Bay.

5._____

6. Estimate the year that the Panama Canal was opened.

6._____

7. Add the year 1850 to the time line to mark the date that California became a state.

7.

8. Name two historical events that occurred after the eighteenth century.

8.

In Exercises 9–12, use the bar graph showing the heights of dams in the United States. (5.2, 5.5)

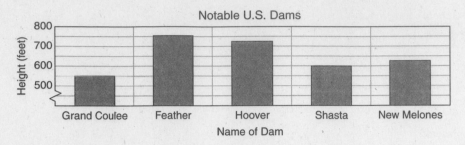

Notable U.S. Dams

9. Without looking at the scale, compare the height of the Grand Coulee Dam with that of the Hoover Dam.

9. _____

10. Estimate the actual difference between the heights of the Grand Coolee Dam and the Hoover Dam.

10. _____

11. Redraw the graph so that it is not misleading. Then explain how you changed the original graph.

11.

12. The states of which dams would more likely use the original graph to attract tourists? Explain your answer.

12.

13. Create a scatter plot for the data (x, y) shown. (5.7)

   (6, 0)   (8, 3)   (10, 7)   (9, 4)   (5, 1)

   (7.5, 2)   (9.5, 5)   (12, 6)   (13, 8)   (11, 10)

13.

#13 to 15
2𝜀

14. Do x and y have a positive correlation, a negative c___ no correlation? (5.7)

15. Based on the scatter plot, estimate the y-coordinate for the x-coordinate value of 15. (5.7)

15. _____

In Exercises 16–21, the list shows the ages of the teaching staff of Frost Middle School's math department. (5.6, 5.8)

21 57 28 45 58 25 35 40 51 32
45 53 32 36 61 27 33 45 28 29

16. Organize the data into a line plot.

16._____

17. How many people are older than 40 years of age?

17._____

18. What is the probability that a student will have a math teacher that is older than 20 and younger than 36 years of age?

18._____

19. What is the probability that a student will have a math teacher that is older than 35 and younger than 50 years of age?

19._____

20. What is the probability that a student will have a math teacher that is older than 49 and younger than 65 years of age?

20._____

21. If the teaching staff remained the same, would your answers to Exercises 18–20 remain the same or change in a year's time? Explain your answer.

21.

In Exercises 22 and 23, the table lists the average number of sit-ups that Judy can complete in one minute each month. (5.4)

Month	September	October	November	December	January	February	March
Sit-ups	6	10	16	25	28	42	38

22. Choose a graph that best represents the data. Then draw the graph and explain your choice.

22._____

23. Write a problem that can be solved by interpreting the graph. Then solve your problem.

23._____

6.2 Short Quiz

Name _____

Date _____

In Exercises 1 and 2, tell whether the numbers are divisible by 2, 3, 4, 5, 6, 8, 9, or 10.

1. 4,292

2. 51,8__

1. _____

2. _____

In Exercises 3 and 4, find all the factors of the number.

3. 84

4. 48

3. _____

4. _____

In Exercises 5 and 6, write the prime factorization of the number in exponent form.

5. 96

6. 216

5. _____

6. _____

7. Write the expanded form and the exponent form of $-56y^2z$.

7. _____

8. Evaluate the expression $-1 \cdot 2^2 \cdot 3^3$.

8. _____

9. You need to arrange 100 chairs in rows for a performance. List three possible arrangements for the chairs.

9. _____

6.4 Short Quiz

Name_____

Date _____

In Exercises 1 and 2, find the greatest common factor of the numbers.

1. 48, 72

1._____

2. $3x^2y$, $12xy^2$

2._____

3. Name a pair of numbers that have a greatest common factor of 15.

3._____

4. Draw a rectangle with an area of 18 square units and with dimensions that are whole numbers. Write the dimensions of the rectangle.

4.

5. What is the perimeter of the rectangle you drew in Exercise 4?

5._____

6. Are the perimeter and area of the rectangle you drew relatively prime? Explain.

6.

7. List the first several multiples of 8 and 12. Use the result to find the least common multiple.

7.

In Exercises 8 and 9, write the prime factorization of each expression. Use the result to find the least common multiple.

8. 36, 54

8.

9. $16xy^2$, $20y^4$

9.

For Exercises 1 and 2, use the divisibility tests to decide whether the number is divisible by 2, 3, 4, 5, 6, 8, 9, and 10. (6.1)

1. 432

1._____

2. 9560

2._____

3. List all of the factors of 240. (6.3)

3._____

In Exercises 4 and 5, write the prime factorization. (6.2)

4. 96

4._____

5. 120

5._____

6. Find the greatest common factor of 15 and 40. (6.3)

6._____

7. Write a pair of numbers whose greatest common factor is 6. (6.3)

7._____

8. Find the least common multiple of 9 and 15. (6.4)

8._____

In Exercises 9–11, simplify the expression. (6.5)

9. $\dfrac{8}{48}$

9._____

10. $\dfrac{72}{132}$

10._____

11. $\dfrac{6x^3}{18x^2y}$

11._____

In Exercises 12 and 13, imagine that you and a friend are walking around a track in the same direction. You start from the same place at the same time. It takes you 6 minutes to walk one lap, and your friend takes 8 minutes to walk one lap. (6.4)

12. After how many minutes will you and your friend be at the same place on the track?

12._____

13. How many laps will you have walked at that time? How many laps will your friend have walked?

13._____

For Exercises 1 and 2, use the divisibility tests to decide whether the number is divisible by 2, 3, 4, 5, 6, 8, 9, and 10. (6.1)

1. 234

2. 5780

3. List all of the factors of 300. (6.2)

In Exercises 4 and 5, write the prime factorization. (6.2)

4. 72

5. 126

6. Write a pair of numbers whose greatest common factor is 24. (6.3)

7. Find the least common multiple of 15 and 20. (6.4)

8. Write a pair of numbers whose least common multiple is 12. (6.4)

In Exercises 9–11, simplify the expression. (6.5)

9. $\dfrac{9}{63}$

10. $\dfrac{96}{176}$

11. $\dfrac{15a^4}{35a^3b}$

In Exercises 12 and 13, imagine that you are laying a tile design using alternating rows of 8-inch and 10-inch tiles. (6.4)

12. After how many inches will the tile design be the same length in each row?

13. How many 8-inch tiles will be in the row at that point? How many 10-inch tiles?

1. _____

2. _____

3. _____

4. _____

5. _____

6. _____

7. _____

8. _____

9. _____

10. _____

11. _____

12. _____

13.

In Exercises 1 and 2, find the greatest common factor of the numerator and denominator. Then use it to simplify the fraction.

1. $\dfrac{8}{20}$

1. _____

2. $\dfrac{36}{96}$

2. _____

In Exercises 3 and 4, simplify the variable expression.

3. $\dfrac{3g^2h}{12gh}$

3. _____

4. $\dfrac{18e^2fg}{32eg^2}$

4. _____

5. Order the fractions $\frac{1}{2}, \frac{1}{4}, \frac{3}{8}, \frac{5}{16}$, and $\frac{21}{32}$ from least to greatest. Then rewrite the list including a fraction of your choice.

5.

6. Write the number $-4\frac{1}{4}$ as a fraction in simplest form.

6. _____

7. Tell whether $\frac{4}{9}$ is rational or irrational. Use a calculator to write the decimal form of the number and tell whether it is terminating, repeating, or non-repeating.

7.

8. Write the decimal 0.375 as a fraction in simplest form.

8. _____

9. Find the perimeter of the figure. Write your result in fraction form, as a mixed number, and as a decimal.

9. _____

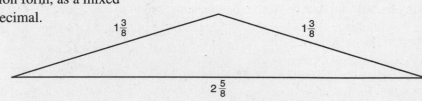

$1\frac{3}{8}$ $1\frac{3}{8}$

$2\frac{5}{8}$

6.8 Short Quiz

Name _____

Date _____

In Exercises 1–4, simplify the expression.

1. $(-6)^3$

1. _____

2. $3x^{-3}$

2. _____

3. $2^5 \cdot 2^{-3}$

3. _____

4. $\dfrac{-3^4}{-3^2}$

4. _____

In Exercises 5 and 6, write the number in scientific notation.

5. 353,000

5. _____

6. 0.00081

6. _____

In Exercises 7 and 8, write the number in decimal form.

7. 4.16×10^6

7. _____

8. 5.49×10^{-3}

8. _____

9. Evaluate $(6.5 \times 10^3)(14 \times 10^{-4})$.

9. _____

10. Write any 8-digit number. Then rewrite it in scientific notation. Tell what power of ten you used to rewrite your number and explain.

10.

In Exercises 1 and 2, use the Divisibility Tests to determine whether the number is divisible by 2, 3, 4, 5, 6, 8, 9, and 10. (6.1)

1. 2484

1._____

2. 7690

2._____

In Exercises 3 and 4, tell whether the number is prime or composite. If it is composite, list all of its factors. (6.2)

3. 71

3._____

4. 92

4._____

5. Decide whether 2448 is divisible by 11, 12, 13, 14, 15, 16, 17, 18, 19, or 20. Use a calculator. (6.1)

5._____

In Exercises 6 and 7, write the prime factorization of the number. (6.2)

6. 204

6._____

7. 175

7._____

In Exercises 8 and 9, write the prime factorization of the expression. (6.2)

8. $24a^2b$

8._____

9. $64y^3$

9._____

Passport to Algebra and Geometry

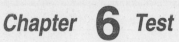
In Exercises 10 and 11, find the greatest common factor and the least common multiple. (6.3, 6.4)

10. 25, 35

10._____

11. $12de^2$, $8e$

11._____

12. There are 52 cards in a regular playing deck. You are playing a card game with friends. How many people can be in the game if all of the cards are to be divided evenly? (6.3)

12._____

In Exercises 13 and 14, simplify the expression. (6.5)

13. $\dfrac{21}{315}$

13._____

14. $\dfrac{3mn}{15n^2}$

14._____

15. Decide which fraction is greater: $\frac{3}{16}$ or $\frac{7}{31}$. (6.5)

15._____

In Exercises 16–18, decide whether the number is rational or irrational. Then write the number in decimal form and state whether the decimal is terminating, repeating, or nonrepeating. (6.6)

16. $\dfrac{-7}{8}$

16._____

17. $\dfrac{5}{18}$

17._____

18. $\sqrt{17}$

18._____

In Exercises 19–25, simplify the expression. (6.7)

19. $(-4)^3$

19._____

20. $(25)^0$

20._____

21. $(-f)^{-4}$

21._____

22. $\dfrac{3^5}{3^3}$

22._____

23. $-5^4 \cdot -5^{-1}$

23._____

24. $4m^5 \cdot 6mn^2$

24._____

25. $\dfrac{27s^3t}{3st^6}$

25._____

26. Write the number 3.4×10^3 in decimal form. (6.8)

26._____

27. Find the product of $(3.5 \times 10^7)(6.8 \times 10^{-2})$ and write it in decimal form and in scientific notation. (6.8)

27._____

28. The sun is about 93 million miles from Earth. Write this number in scientific notation. (6.8)

28._____

29. Describe the pattern. Then draw and write the next three terms in the sequence. (6.9)

29.

30. Draw 3 different rectangles with an area of 28 square units and with dimensions that are whole numbers. Give the dimensions of each. Then tell their perimeters. (6.4)

30.

Form B
(Page 1 of 3 pages)

Name_____

Date _____

In Exercises 1 and 2, by which numbers is the given number divisible? (6.1)

1. 8240
   a. 2, 3, 4, 5, and 10
   b. 5 and 10
   c. 2, 4, 5, 8, and 10
   d. 2, 4, 5, 8, and 9

1._____

2. 8163
   a. 3, 6, and 9
   b. 3 and 9
   c. 3, 6, 8, and 9
   d. 3, 7, and 9

2._____

In Exercises 3 and 4, is the number prime or composite? What are its factors? (6.2)

3. 87
   a. prime
   b. composite; factors 1, 3, 7, 9, 15, 27, 29
   c. composite; factors 1, 3, 29
   d. composite, factors 1, 3, 27

3._____

4. 89
   a. prime
   b. composite; factors 1, 3, 7, 9, 15, 29
   c. composite; factors 1, 7, 17
   d. composite; factors 1, 7, 9, 13, 17, 23

4._____

In Exercises 5–8, what is the prime factorization? (6.2)

5. 225
   a. $5^2 \cdot 3^2$
   b. $25 \cdot 9$
   c. $45 \cdot 5$
   d. $5^3 \cdot 2$

5._____

6. 504
   a. $56 \cdot 9$
   b. $3^2 \cdot 7^2 \cdot 2^2$
   c. $63 \cdot 8$
   d. $2^3 \cdot 7 \cdot 3^2$

6._____

7. $32r^2s$
   a. $4 \cdot 8 \cdot r \cdot r \cdot s$
   b. $2^5 \cdot r \cdot r \cdot s$
   c. $2^5 \cdot 2 \cdot r \cdot s$
   d. $32 \cdot r \cdot s$

7._____

8. $81w^2xy^3$
   a. $9 \cdot 9 \cdot w \cdot x \cdot y$
   b. $3^4 \cdot w^2 \cdot y^3$
   c. $3^4 \cdot w \cdot w \cdot x \cdot y \cdot y \cdot y$
   d. $81 \cdot w \cdot w \cdot x \cdot y \cdot y \cdot y$

8._____

In Exercises 9 and 10, what is the greatest common factor? (6.3)

9. 42 and 56

    a. 8   **b.** 7   **c.** 14   **d.** 2

9._____

10. $12g^2h^3i$ and $18gj^2$

    a. $6g^3h^3ij^2$ **b.** $6g^2h$ **c.** $6gh$ **d.** $6g$

10._____

In Exercises 11 and 12, what is the least common multiple? (6.4)

11. 36 and 48

    a. 48   **b.** 96   **c.** 144   **d.** 124

11._____

12. $9k^2lm$ and $15km^2$

    a. $3k^2lm^2$ **b.** $45k^2lm^2$ **c.** $135k^3lm^3$ **d.** $30k^3lm^2$

12._____

In Exercises 13 and 14, what is the simplified form of the expression? (6.5)

13. $\dfrac{76}{120}$

    a. $\dfrac{38}{60}$ **b.** $\dfrac{1}{3}$ **c.** $\dfrac{19}{30}$ **d.** $\dfrac{9}{15}$

13._____

14. $\dfrac{6th^2e}{21t^2h^2}$

    a. $\dfrac{2}{7}te$ **b.** $\dfrac{2e}{7t^2}$ **c.** $\dfrac{3e}{7t}$ **d.** $\dfrac{6e}{21}$

14._____

15. Which is the order from least to greatest for the fractions $\frac{1}{4}, \frac{3}{8}, \frac{7}{12}$, and $\frac{5}{16}$? (6.5)

    a. $\frac{1}{4}, \frac{3}{8}, \frac{7}{12}, \frac{5}{16}$ **b.** $\frac{1}{4}, \frac{3}{8}, \frac{5}{16}, \frac{7}{12}$

    c. $\frac{1}{4}, \frac{5}{16}, \frac{3}{8}, \frac{7}{12}$ **d.** $\frac{7}{12}, \frac{3}{8}, \frac{1}{4}, \frac{5}{16}$

15._____

16. Is the fraction $\frac{7}{3}$ rational or irrational? Is the number in decimal form terminating, repeating, or nonrepeating? (6.6)

    a. rational; repeating   **b.** irrational; repeating

    c. rational; terminating   **d.** irrational; nonrepeating

16._____

In Exercises 17–21, what is the simplified form of the expression? (6.5, 6.7)

17. xy^{-3} 17._____

a. $\dfrac{1}{xy^3}$ b. $\dfrac{x}{y^3}$ c. xy^3 d. $\dfrac{1}{xy^{-3}}$

18. $\dfrac{2^8}{2^4}$ 18._____

a. 4   b. $\dfrac{1}{4}$ c. 16   d. $\dfrac{1}{16}$

19. $-3^2 \cdot -3^{-3}$ 19._____

a. 3   b. -3 c. $-\dfrac{1}{3}$ d. $\dfrac{1}{3}$

20. $\dfrac{36h^2m}{16hm^4}$ 20._____

a. $\dfrac{9h}{4m^3}$ b. $\dfrac{9h^2}{4m^4}$ c. $\dfrac{h}{m^3}$ d. $\dfrac{36h}{16m^3}$

21. $16v^3u \cdot 5v^4u^2w$ 21._____

a. $80v^3u^3w$ b. $80v^7u^3w$ c. $80v^{12}u^2w$ d. $80v^7u^2w$

22. How is 4.37×10^8 written in standard form? (6.8)   22._____

a. 437,000,000   b. 43,700,000,000
c. 437,108   d. 43,700,000

23. How is 0.000872 written in scientific notation? (6.8)   23._____

a. 872×10^4 b. 8.72×10^{-4}
c. 8.72×10^4 d. 872×10^{-4}

24. What is the product of (7×10^{-3}) and (4.3×10^9)? (6.8)   24._____

a. 30.1×10^3 b. 3,010,000   c. 301,000   d. 30,100,000

25. A rectangle has an area of 24 inches. Which dimensions for the rectangle are NOT possible? (6.4)   25._____

a. 4 inches by 6 inches   b. 12 inches by 2 inches
c. 3 inches by 8 inches   d. 10 inches by 4 inches

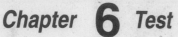
In Exercises 1–3, use the number 23,87 $\boxed{?}$. (6.1)

1. Which digit would make the number divisible by 9?

    1. _____

2. Which digit would make the number divisible by 8?

    2. _____

3. If you used the last digit you found in Exercise 2, by which other digits would the completed number be divisible?

    3. _____

4. State a rule for divisibility by 15. (6.1)

    4.

5. A box has a volume of 30 cubic units (Volume = length × width × height). List the factors of 30. Use the list to identify three possible dimensions for the length, width, and height of the box. (6.2)

    5.

**In Exercises 6 and 7, draw a tree diagram to show the prime factorization.
Then write the prime factorization in exponent form. (6.2)**

6. 60

    6. _____

7. 192

    7. _____

In Exercises 8 and 9, write the expanded form and the exponent form of the expression. (6.2)

8. $32ab^4c^5$

    8. _____

9. $-54t^2uv^3$

    9. _____

In Exercises 10–12, find the greatest common factor and the least common multiple. (6.3, 6.4)

10. 15, 40

10._____

11. $2x^2y$, $6xy^2$

11._____

12. $10ab^2c^2$, $45abc^3$

12._____

13. Write two fractions that are between $\frac{1}{2}$ and $\frac{5}{8}$. Explain how you found the fractions you chose.

13._____

In Exercises 14–18, simplify the variable expression. (6.5, 6.7)

14. $x^{-2} \cdot x^8$

14._____

15. $r^4s^3 \cdot r^{-5}s^2 \cdot r^{-2}s^{-3}$

15._____

16. $\dfrac{-15a^2b^2}{3ab^4}$

16._____

17. $3^2g^4hi^{-2} \cdot 2^3gh^4$

17._____

18. $\dfrac{42c^5de^2}{-9c^2d^3}$

18._____

19. Write 0.4444 . . . as a fraction. Simplify the result. (6.6)

19._____

20. Find the product of (3.5×10^2) and (14.23×10^{-7}). Write it in decimal form and in scientific notation. (6.8)

20._____

In Exercises 21–23, tell whether the number is rational or irrational. Then write the decimal form of the number and state whether the decimal is terminating, repeating, or nonrepeating. (6.6)

21. $\frac{1}{12}$

21._____

22. $\sqrt{225}$

22._____

23. $\sqrt{50}$

23._____

24. Describe the pattern. Then draw the next three terms in the sequence. (6.9)

24.

In Exercises 25–27, the speed of sound varies with the material with which it is traveling through. For each material, write the speed of sound in standard form. Then find the speed of sound in feet per minute. Write your result using scientific notation. (6.8) *(Source: 1997 World Almanac)*

25. Ice cold vapor: 4.708×10^3 feet per second

25._____

26. Hardwood: 1.262×10^4 feet per second

26._____

27. Brick: 1.196×10^4 feet per second

27._____

28. Order the speeds from slowest to fastest. (6.8)

28._____

In Exercises 1 and 2, describe the pattern and draw the next two figures. Then write an expression for the perimeter of the regular polygon and evaluate the expression when $x = 5$. (1.7, 2.2)

1.

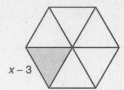 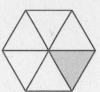

$x - 3$

1. _____

2.

$2x + 5$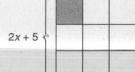

2. _____

In Exercises 3–9, use a calculator to evaluate the expression. Round to two decimal places if necessary. (1.3, 1.4)

3. 8^5

3. _____

4. $\left(\dfrac{5}{8}\right)^3$

4. _____

5. $\sqrt{65}$

5. _____

6. $17 + (4^3 - 19) \div 3$

6. _____

7. $\sqrt{313.26}$

7. _____

8. $(4.7)^4$

8. _____

9. $4(33 - 19) + 3^5$

9. _____

10. Apply the distributive property to an algebraic expression of your choice. Evaluate the expression when the value of your variable is 5. (2.1)

10.

11. Tell what operations you would use to solve the equation $3x + 4 = 28$. (4.1)

11. _____

In Exercises 12–17, evaluate the expression. (1.4, 3.1–3.6)

12. $(3)(-5)(10)(-6)(-1)$

12. _____

13. $-|7| - |-7| - 3 + |-3|$

13. _____

14. $|-17| - |17| + 9 - 8$

14. _____

15. $2^4 - (9 - 3) \cdot 2$

15. _____

16. $(4 - 11)^2 \div 7 + 3$

16. _____

17. $\dfrac{-156}{12}$

17. _____

In Exercises 18–23, write an algebraic equation or inequality for the sentence. Then solve. (2.3–2.5, 2.7, 2.9, 3.7)

18. The difference of w and 4 is -5.

18. _____

19. 9 is the quotient of 108 and t.

19. _____

20. 29 increased by f is -2.

20. _____

21. The product of e and 14 is greater than -126.

21. _____

22. -3 is greater than or equal to the sum of 5 and z.

22. _____

23. 7 is greater than or equal to the quotient of b and 16.

23. _____

In Exercises 24–28, write the expression without parenthesis and combine like terms when possible. Then evaluate it when $a = -3$, $b = 4$, and $c = 10$. (3.1–3.5)

24. $4(b - a + c)$

24. _____

25. $-5(a - 3b + c)$

25. _____

26. $|-b| + 4(a + c)$

26. _____

27. $5 + -2(b + c) + 3c$

27. _____

28. $-3a + 9b - 4c + 2a - 5b$

28. _____

In Exercises 29–31, plot the points on the coordinate plane and identify the quadrant in which the point lies. (3.8)

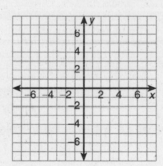

29. $A(-3, -3)$

29. _____

30. $B(-3, 4)$

30. _____

31. $C(4, 4)$

31. _____

32. Add a fourth point (D) to the coordinate plane above so that the points can be connected to form a square. Write the coordinates of the point and the quadrant in which it lies. (3.8)

32. _____

33. Find the perimeter and the area of the square formed by Exercises 29–32. (3.8)

33. _____

In Exercises 34–36, write the reciprocal. (4.3)

34. 1

34._____

35. 5

35._____

36. $-\dfrac{1}{7}$

36._____

In Exercises 37–40, solve the equation. Then check your solution. (4.1, 4.2)

37. $5t - 17 = 68$

37._____

38. $3v - 8 + 4v = 27$

38._____

39. $\dfrac{r}{8} + 23 = 38$

39._____

40. $\dfrac{3}{5}g + 9 - \dfrac{2}{5}g = 30$

40._____

In Exercises 41–43, the sum of the measures of the angles in a trapezoid is 360°. (4.4)

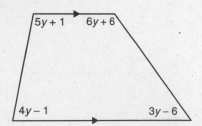

41. Write an equation for the sum of the measures of the angles of the trapezoid shown.

41._____

42. Find the value of y.

42._____

43. Write the measure of each angle of the trapezoid.

43._____

In Exercises 44–49, solve the equation. (4.4, 4.5)

44. $27 + 4b = 13b$

44. _____

45. $7r = 3r - 52$

45. _____

46. $9s + 7 = 4s - 8$

46. _____

47. $5(b - 3) = 7b - 14$

47. _____

48. $6(e + 1) = 4(e + 2)$

48. _____

49. $\frac{1}{3}(h + 6) = \frac{5}{6}h$

49. _____

In Exercises 50 and 51, solve the equation. Round the result to 2 decimal places. (4.7)

50. $-25x + 17 = 12x - 8$

50. _____

51. $41.6j - 15.2 = 7.1(2.3j - 8)$

51. _____

In Exercises 52 and 53, a rectangle is 8 inches long and has an area of 96 inches. (4.8)

52. Find the width and the perimeter of the rectangle.

52. _____

53. Draw a rectangle with the same area and a different perimeter. Write the dimensions and the perimeter of the new rectangle.

53. _____

In Exercises 54–57, use the scatter plot which compares the height and age of U.S. children. **(5.7)** *(Source: Physicians Handbook, c/o World Almanac)*

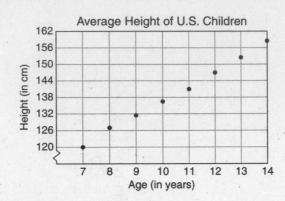

Average Height of U.S. Children

54. Do the age and height have a positive correlation, a negative correlation, or no correlation? Explain your reasoning.

54.

55. Write a sentence that describes the relationship between age and height of U.S. children.

55.

56. Estimate the height at age 12.

56. _____

57. Predict the height of the average 15 year old, based on the graph.

57. _____

In Exercises 58–60, the table lists the scores that the Latter twins got on their math tests over a 6-week period. **(5.4)**

Name	Week 1	Week 2	Week 3	Week 4	Week 5	Week 6
Todd	85	93	87	92	90	100
Greg	82	85	87	90	93	98

58. Represent this data graphically. Explain your choice of graph.

58.

59. Which twin made steady progress throughout the six weeks?

59. _____

60. Predict Todd's and Greg's scores for Week 7. Explain your predictions.

60.

In Exercises 61–65, find the probability that the spinner will land on the given letter. (5.8)

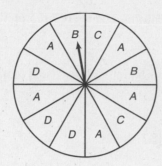

61. A

61. _____

62. B

62. _____

63. C

63. _____

64. D

64. _____

65. Not A

65. _____

In Exercises 66–69, find the greatest common factor and the least common multiple. (6.2–6.4)

66. 36, 6

66. _____

67. 120, 420

67. _____

68. $4x^2y^3$, $18xy^4$

68. _____

69. $16a^3b$, $12a^2b^5$

69. _____

In Exercises 70–72, simplify the fraction. Then write it in decimal form. Round your answer to 3 decimal places, if necessary. (6.5, 6.6)

70. $\dfrac{21}{24}$

70. _____

71. $\dfrac{48}{144}$

71. _____

72. $\dfrac{96}{99}$

72. _____

73. Use a calculator to find the value of $\sqrt{54}$. Round your answer to 3 decimal places if necessary. (6.6)

73. _____

74. Write two expressions whose greatest common factor is $7ab$. (4.3)

74._____

75. Write two expressions whose least common multiple is $20c^2$. (4.4)

75._____

In Exercises 76–82, simplify the expression. (6.7)

76. 3^{-3}

76._____

77. 15^0

77._____

78. x^{-4}

78._____

79. $6^{-2} \cdot 6^5$

79._____

80. $\dfrac{y^4}{y^3}$

80._____

81. $14b \cdot 3b^3$

81._____

82. $\dfrac{25e^4}{15e^5}$

82._____

In Exercises 83 and 84, the National Fire Protection Association documents that public fire departments responded to 1,965,500 fires in 1995. (6.8)

(Source: 1997 World Almanac)

83. Write this number in scientific notation.

83._____

84. Find the average number of fires per month in 1995. Express your answer in standard form and in scientific notation.

84._____

85. Write 0.0000564 in scientific notation.

85._____

Passport to Algebra and Geometry

7.2 Short Quiz

Name_____

Date _____

In Exercises 1–4, add or subtract. Then simplify if possible.

1. $-\dfrac{5}{7} + \dfrac{2}{7}$

1._____

2. $4\dfrac{4}{9} - 2\dfrac{1}{9}$

2._____

3. $\dfrac{6y}{12} + \dfrac{5y}{12}$

3._____

4. $\dfrac{3}{4r} - \dfrac{6}{4r}$

4._____

In Exercises 5 and 6, solve the equation. Then simplify if possible.

5. $\dfrac{5}{11} + f = -\dfrac{4}{11}$

5._____

6. $u - \dfrac{5}{6} = \dfrac{1}{6}$

6._____

In Exercises 7–10, find the sum or difference. Then simplify, if possible.

7. $\dfrac{3}{4} + \dfrac{5}{6} - \dfrac{1}{2}$

7._____

8. $-\dfrac{3}{8y} + \dfrac{5}{12y}$

8._____

9. $\dfrac{5}{st} + \dfrac{6}{s}$

9._____

10. $\dfrac{6}{c} + \dfrac{8}{18}$

10._____

11. Write a number sentence comparing two unlike fractions. Then find the difference and simplify if possible.

11._____

Name_____

Date _____

In Exercises 1 and 2, evaluate the expression by first rewriting in decimal form. Round your result to two decimal places.

1. $\dfrac{8}{17}x - \dfrac{7}{16}x$

   1._____

2. $\dfrac{3}{8} + \dfrac{4}{7} + \dfrac{9}{16} - 1$

   2._____

3. Write three numbers that, rounded to two decimal places, would have a value of 2.89.

   3.

In Exercises 4–7, multiply. Then simplify, if possible.

4. $\dfrac{3}{4} \cdot \left(-\dfrac{4}{7}\right) \cdot \dfrac{21}{25}$

   4._____

5. $3\dfrac{5}{6} \cdot \left(-2\dfrac{7}{8}\right)$

   5._____

6. $\dfrac{2w}{3} \cdot \dfrac{12}{5}$

   6._____

7. $-\dfrac{15a^3}{a^2} \cdot \dfrac{a}{10a^2}$

   7._____

8. Use a calculator to find the area of the triangle. Round your result to three decimal places.

   8._____

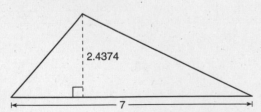

2.4374

7

(Use after Lesson 7.4)

Name_____

Date_____

In Exercises 1 and 2, find the sum or difference and simplify, if possible. (7.1, 7.2)

1. $\frac{9}{16} - \frac{3}{24}$

1._____

2. $\frac{4}{15} + \frac{62}{75}$

2._____

In Exercises 3 and 4, find the product and simplify, if possible. (7.4)

3. $\frac{5}{8} \cdot \left(-\frac{3}{4}\right) \cdot \frac{1}{3}$

3._____

4. $\frac{6a}{2} \cdot \left(-\frac{2}{3}\right)$

4._____

In Exercises 5 and 6, solve the equation. (7.2)

5. $\frac{3}{8} + d = \frac{5}{12}$

5._____

6. $x - \frac{1}{4} = \frac{5}{2}$

6._____

7. Draw a rectangle that has a perimeter of $8\frac{1}{2}$. Find its area. (7.1, 7.2, 7.4)

7._____

8. Evaluate the expression by first rewriting in decimal form. Round your results to two decimal places. (7.3)

$1\frac{3}{5} - 2\frac{5}{6} - \frac{14}{15} + 2\frac{7}{11}$

8._____

9. Write the decimals as fractions and multiply. (7.4)

$0.\overline{3}$; 0.2

9._____

10. You find one pair of binoculars originally priced at $48.95 but marked "$\frac{1}{3}$ off," and another pair of binoculars originally priced at $69.95 but marked "one-half off." Which pair of binoculars is less expensive? By how much? (7.4)

10._____

In Exercises 1 and 2, find the sum or difference and simplify, if possible. (7.1, 7.2)

1. $\frac{8}{15} - \frac{9}{20}$

1. _____

2. $\frac{3}{14} + \frac{56}{70}$

2. _____

In Exercises 3 and 4, find the product and simplify, if possible. (7.4)

3. $\frac{9}{10} \cdot \left(-\frac{2}{7}\right) \cdot \frac{3}{5}$

3. _____

4. $\frac{4n}{3} \cdot \left(-\frac{2}{6}\right)$

4. _____

In Exercises 5 and 6, solve the equation. (7.2)

5. $\frac{5}{9} + d = \frac{4}{15}$

5. _____

6. $k - \frac{3}{4} = -\frac{7}{12}$

6. _____

7. Draw a rectangle that has a perimeter of $6\frac{1}{2}$. Find its area. (7.1, 7.2, 7.4)

7. _____

8. Evaluate the expression by first rewriting in decimal form. Round your result to two decimal places. (7.3)

$2\frac{5}{9} + 3\frac{3}{7} - 1\frac{7}{10} + 4\frac{1}{8}$

8. _____

9. Write the decimals as fractions and multiply. (7.4)
   $0.375;\ 0.\overline{6}$

9. _____

10. One store is selling running shoes originally priced at $54.99 a pair but marked "$\frac{1}{5}$ off." Another store is selling the shoes at the same price, but takes $\frac{1}{20}$ off the price of the first pair and $\frac{1}{4}$ off the price of the second pair. Which is a better deal if you are buying two pairs of shoes? By how much? (7.4)

10. _____

Name _____

Date _____

In Exercises 1–4, simplify the expression.

1. $\dfrac{2}{3} \div \dfrac{4}{9}$ 1. _____

2. $\dfrac{3}{8} \div -1\dfrac{2}{5}$ 2. _____

3. $-\dfrac{2}{5} \div \dfrac{10}{x}$ 3. _____

4. $\dfrac{12m}{5} \div \dfrac{6m}{20}$ 4. _____

5. You have $4\dfrac{1}{2}$ chocolate bars. You need $\dfrac{1}{4}$ of a bar to make one s'more.   5. _____
   How many s'mores can you make?

6. Write the percent of the figure that is   6. _____
   shaded.

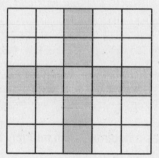

In Exercises 7 and 8, write each portion as a percent.

7. $\dfrac{23}{50}$ 7. _____

8. $\dfrac{64}{400}$ 8. _____

In Exercises 9 and 10, choose a fraction between $\dfrac{1}{2}$ and 1.

9. Write the fraction as a percent.   9. _____

10. Draw a geometric model to show the percent.   10. _____

Name_____

Date _____

In Exercises 1 and 2, rewrite the percent as a decimal.

1. 43.5%

1._____

2. $27\frac{1}{2}\%$

2._____

In Exercises 3 and 4, rewrite the decimal as a percent.

3. 0.843

3._____

4. 1.67

4._____

In Exercises 5 and 6, rewrite the percent as a fraction in simplest form.

5. 28%

5._____

6. 110%

6._____

In Exercises 7 and 8, write the percent as a decimal. Then multiply to find the percent of the number.

7. 36% of 36

7._____

8. 125% of 160

8._____

9. Draw a rectangle and identify its dimensions. Then draw another rectangle whose dimensions are 50% of the dimensions of the first rectangle. Identify its dimensions.

9._____

10. Find the perimeter of the rectangles you drew in Exercise 9. Is the perimeter of the smaller rectangle 50% of the perimeter of the larger rectangle? Explain.

10._____

In Exercises 1–4, find the sum or difference and simplify, if possible. (7.1, 7.2)

1. $\dfrac{3}{11} + \dfrac{1}{2}$

2. $\dfrac{15}{16} - \dfrac{3}{4}$

3. $\dfrac{6}{5}t - \dfrac{4}{5}t$

4. $-\dfrac{4}{ab} + \dfrac{9}{b}$

1._____

2._____

3._____

4._____

In Exercises 5–8, find the product or quotient and simplify, if possible. (7.4, 7.5)

5. $2\dfrac{3}{8} \cdot \dfrac{4}{9}$

6. $6 \div \dfrac{1}{2}$

7. $12 \cdot \dfrac{5m}{8}$

8. $-\dfrac{5}{d} \div \dfrac{15}{d}$

5._____

6._____

7._____

8._____

In Exercises 9–14, solve the equation. (7.1–7.5)

9. $z + \dfrac{7}{8} = \dfrac{3}{8}$

10. $3t + \dfrac{1}{15} = \dfrac{4}{5}$

11. $b - \dfrac{4}{7} = \dfrac{25}{28}$

12. $17u - \dfrac{1}{5} = 9u$

13. $\dfrac{1}{10}q = \dfrac{3}{14}$

14. $\dfrac{15}{11} = -\dfrac{1}{2}t - \dfrac{3}{22}$

9._____

10._____

11._____

12._____

13._____

14._____

15. Find the perimeter of the hexagon. (7.2)

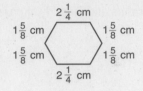

15. _____

16. Find the perimeter and the area of the region. (7.2, 7.4)

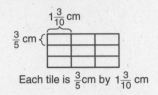

Each tile is $\frac{3}{5}$ cm by $1\frac{3}{10}$ cm

16. _____

In Exercises 17–20, use a calculator to evaluate the expression. Round your result to two decimal places. (7.4)

17. $\frac{15}{23} + \frac{4}{9}$

17. _____

18. $\frac{34}{81}y - \frac{56}{67}y$

18. _____

19. $\frac{12}{13} \cdot \frac{15}{}$

19. _____

20. $-\frac{}{41} \div \frac{15}{19}$

20. _____

21. Shade in sections of the grid. Write the fraction that represents the portion of the grid's area that you shaded. Then write the fraction as a percent. (7.6)

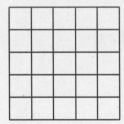

21. _____

22. Rewrite 0.042 as a percent. (7.8)

22. _____

23. Rewrite 146% as a fraction. Then simplify. (7.7)

23. _____

In Exercises 24 and 25, find the percent of the number. (7.8)

24. 46% of 50

24. _____

25. 250% of 66

25. _____

26. Find the total area of the figure. Then find the percent of the total area that is shaded, the percent that is dotted, and the percent that is striped. (7.8)

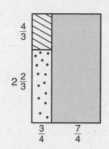

26._____

In Exercises 27 and 28, use the circle graph at the right, which shows Evan's typical day. Round your answers to the nearest hour. (7.8)

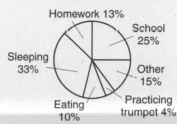

27. How much time does Evan spend on school and homework combined?

27._____

28. Compare the number of hours Evan spends on two activities during the course of his day.

28._____

29. You stop in a restaurant for lunch. Your lunch costs $9.45, including tax. You leave a tip of 15%. What is the total cost of your lunch out? (7.9)

29._____

In Exercises 30–33, use the following information. Round to one decimal place.

Your parents are selling their collection of old records. Of the 110 records in the collection, 65 are rock and roll, 28 are soul, and 17 are folk music. Of the 65 rock and roll records, 18 are by the Beatles and 8 are by the Rolling Stones. (7.9)

30. What percent of the total number of records are soul records?

30._____

31. What percent of the total number of rock and roll records are by the Beatles?

31._____

32. What percent of the total record collection are by the Beatles?

32._____

33. Write another percent problem about the record collection. Solve the problem.

33._____

In Exercises 1–4, find the sum or difference and simplify. (7.1, 7.2)

1. $\frac{4}{9} + \frac{2}{9}$

   a. $\frac{6}{18}$ b. $\frac{1}{3}$ c. $\frac{6}{9}$ d. $\frac{2}{3}$

   1._____

2. $\frac{11}{16}u - \frac{9}{16}u$

   a. $\frac{2}{0}u$ b. $\frac{20}{32}u$ c. $\frac{1}{8}u$ d. $\frac{2}{16}u$

   2._____

3. $\frac{3}{8} + \frac{5}{16}$

   a. $\frac{11}{16}$ b. $\frac{8}{16}$ c. $\frac{11}{24}$ d. $\frac{3}{8}$

   3._____

4. $\frac{11}{20} - \frac{2}{15}$

   a. $\frac{9}{5}$ b. $\frac{5}{12}$ c. $\frac{9}{20}$ d. $\frac{25}{60}$

   4._____

In Exercises 5–8, find the product or quotient and simplify.

5. $\frac{5}{8} \cdot \frac{16}{25}$

   a. $\frac{2}{5}$ b. $\frac{70}{200}$ c. $\frac{2}{50}$ d. $\frac{35}{100}$

   5._____

6. $\frac{2}{5} \div \frac{7}{10}$

   a. $\frac{2}{7}$ b. $\frac{4}{7}$ c. $\frac{14}{50}$ d. $\frac{7}{25}$

   6._____

7. $2\frac{4}{5} \times 3\frac{1}{2}$

   a. $\frac{4}{5}$ b. $9\frac{4}{5}$ c. $6\frac{4}{10}$ d. 49

   7._____

8. $-\frac{4e}{9} \div \frac{16e}{3}$

   a. $-\frac{1}{12}$ b. $-\frac{e}{12}$ c. $-2\frac{10}{27}$ d. $-\frac{64}{27}$

   8._____

9. What is the perimeter of the pentagon? (7.2)

   a. $25\frac{1}{12}$ cm

   b. $23\frac{1}{12}$ cm

   c. $24\frac{15}{12}$ cm

   d. $25\frac{1}{4}$ cm

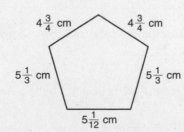

   9._____

In Exercises 10–12, solve the equation. (7.2, 7.4, 7.5)

10. $g + \frac{3}{5} = \frac{14}{15}$

    a. $g = \frac{11}{15}$ **b.** $g = \frac{11}{5}$ **c.** $g = \frac{1}{3}$ **d.** $g = \frac{2}{3}$

10._____

11. $6q - \frac{7}{3} = \frac{5}{6}$

    a. $q = \frac{1}{3}$ **b.** $q = \frac{19}{36}$ **c.** $q = \frac{19}{6}$ **d.** $q = \frac{35}{18}$

11._____

12. $\frac{3}{8}h = 2\frac{1}{2}$

    a. $h = \frac{8}{3}$ **b.** $h = \frac{15}{16}$ **c.** $h = 3\frac{1}{3}$ **d.** $h = \frac{20}{3}$

12._____

In Exercises 13 and 14, use the rectangle. (7.2, 7.4)

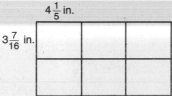

$4\frac{1}{5}$ in.

$3\frac{7}{16}$ in.

Each tile is $4\frac{1}{5}$ in. $\times$ $3\frac{7}{16}$ in.

13. What is the perimeter?

    a. $18\frac{19}{40}$ in.   **b.** $18\frac{5}{4}$ in.   **c.** $38\frac{19}{20}$ in.   **d.** $39\frac{9}{10}$ in.

13._____

14. What is the area?

    a. $86\frac{5}{8}$ in.^{2} **b.** $72\frac{21}{40}$ in.^{2} **c.** $72\frac{1}{2}$ in.^{2} **d.** $86\frac{1}{2}$ in.^{2}

14._____

15. How is $\frac{13}{20}$ expressed as a percent? (7.6)

    a. 13%   **b.** 65%   **c.** 6.5%   **d.** $\frac{13}{20}$%

15._____

16. How is 0.417 expressed as a percent? (7.7)

    a. 0.417%   **b.** 4.17%   **c.** 41.7%   **d.** 417%

16._____

17. How is 87.9% written as a decimal? (7.7)

    a. 8790   **b.** 87.9   **c.** 8.79   **d.** 0.879

17._____

18. How is 48% expressed as a fraction in simplest form? (7.7)

    a. $\frac{48}{100}$ **b.** $\frac{24}{50}$ **c.** $\frac{12}{25}$ **d.** $\frac{1}{2}$

18._____

19. What is 35% of 200? (7.8)   19._____

   a. 35   **b.** 70   **c.** 140   **d.** 700

20. 28 is what percent of 140? (7.8)   20._____

   a. 28%   **b.** 50%   **c.** 25%   **d.** 20%

In Exercises 21–23, use the figure at the right. (7.8)

$\frac{5}{4}$ in.

$\frac{3}{4}$ in.

$\frac{7}{5}$ in. $\frac{3}{5}$ in.

21. What is the total area of the figure?   21._____

   a. 4 in.^{2} **b.** $\frac{3}{5}$ in.^{2} **c.** 2 in.^{2} **d.** 1 in.^{2}

22. What percent of the total area is white?   22._____

   a. ≈ 20%   **b.** ≈ 18.8%   **c.** ≈ 19.8%   **d.** ≈ 25%

23. What percent of the total area is starred?   23._____

   a. 75%   **b.** 65%   **c.** 78%   **d.** 70%

In Exercises 24–26, use the following information.

You take a survey to determine the amount of time students in your math class study for a test. The results are: none (8.5%), 15 minutes or less (12%), between 15 and 30 minutes (37.5%), between 30 minutes and an hour (32%), more than an hour (10%). (7.9)

24. Of 56 students, how many would you expect to study for between 15 and 30 minutes?   24._____

   a. 20   **b.** 21   **c.** 19   **d.** 25

25. Of 110 students, about how many would you expect to study for between 30 minutes and an hour?   25._____

   a. 30   **b.** 33   **c.** 38   **d.** 35

26. Of 35 students, about how many would you expect to study for 30 minutes or more?   26._____

   a. 14   **b.** 12   **c.** 15   **d.** 11

In Exercises 1–4, find the sum or difference and simplify, if possible. **(7.2)**

1. $\frac{1}{5} + \frac{1}{4} + \frac{1}{8}$

1._____

2. $\frac{8}{9} - \frac{15}{23}$

2._____

3. $\frac{6}{ab} - \frac{3}{b}$

3._____

4. $-\frac{5}{8} - \frac{6}{fg}$

4._____

In Exercises 5–8, find the product or quotient and simplify, if possible.
(7.4, 7.5)

5. $\frac{1}{2} \cdot \frac{3}{4} \cdot 8\frac{2}{5}$

5._____

6. $8\frac{3}{4} \div 2\frac{5}{7}$

6._____

7. $\frac{g}{2} \cdot \frac{2f}{g} \cdot \frac{1}{f}$

7._____

8. $\frac{x^2}{5} \div \left(-\frac{x}{10}\right)$

8._____

In Exercises 9–12, solve the equation. **(7.2, 7.5)**

9. $x - \frac{3}{16} = \frac{1}{8}$

_____

10. $25j - \frac{4}{7} = \frac{16}{21}$

_____

11. $12y + 6 = -12$

11._____

12. $\frac{18}{9} = -\frac{2}{3}w - \frac{5}{12}$

12._____

13. Find the perimeter and area of the region. **(7.2, 7.4)**

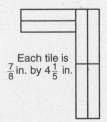

Each tile is
$\frac{7}{8}$ in. by $4\frac{1}{5}$ in.

13._____

In Exercises 14 and 15, write an equation for the verbal model. Then solve. (7.2, 7.4, 7.5)

14. The product of 15 and t increased by $\frac{4}{5}$ is $-\frac{9}{10}$.

14._____

15. The quotient of v and 30 is $\frac{2}{3}$.

15._____

In Exercises 16 and 17, use a calculator to evaluate the expression. Round your result to two decimal places. (7.3)

16. $\dfrac{37}{45t} \cdot \left(-\dfrac{8t}{15}\right)$

16._____

17. $\dfrac{21}{34} \div \dfrac{42}{17}$

17._____

In Exercises 18–21, rewrite as a percent. Round your results to two decimal places. (7.6, 7.7)

18. $\dfrac{15}{38}$

18._____

19. $2\frac{3}{7}$

19._____

20. 0.589

20._____

21. 0.0734

21._____

22. Write 2.7% as a decimal. (7.7)

22._____

23. Write a fraction or mixed number in simplest form. Then write its decimal and percent equivalents. (7.7)

23.

24. Find 35% of 87. (7.8)

24._____

25. 17.2 is what percent of 25? (7.8)

25._____

26. Find the total area of the figure. Then find the percent of the total area that is white, the percent that is shaded, the percent that is starred, and the percent that is striped. (7.8)

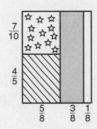

26._____

27. Shade in sections of the grid. Write the fraction that represents the portion of the grid's area that is not shaded. Then write the fraction as a decimal and a percent. (7.6, 7.7)

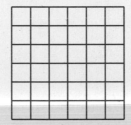

27._____

In Exercises 28–30, use the following information.

Your class is having a plant sale to raise funds for a field trip. You have 85 sunflowers to sell. Of these, 38 will have orange flowers. Of these 38 orange sunflowers, 15 will grow over six feet tall and 18 will grow less than four feet tall. (7.9)

28. What percent of the total number of sunflowers are *not* orange?

28._____

29. What percent of the total number of orange sunflowers will grow less than four feet?

29._____

30. 20% of the sunflowers grow less than four feet tall. How many sunflowers is this?

30._____

In Exercises 31–33, use the following information.

The class goes on its field trip to Washington, D.C. Students are surveyed to determine the sights they would most like to see. The results are: The White House (28.2%), the Washington Monument (23.8%), the Air and Space Museum (30.3%), the Jefferson Memorial (8.9%), and no preference (8.8%). (7.9)

31. Of 138 students, how many would you expect to visit the White House first?

31._____

32. Of 325 students, how many would you expect to visit the Air and Space museum first?

32._____

Name_____

Date _____

1. Is 85 chairs/17 tables a ratio or a rate? Explain.

1._____

In Exercises 2 and 3, write each quotient as a ratio. Then simplify.

2. 660 feet/1 mile

2._____

3. 1 kilogram/500 grams

3._____

4. A 10-pound bag of ice sells for $0.79. Determine your own price for a 15-pound bag of ice that would be a better bargain.

4._____

In Exercises 5 and 6, solve the proportion.

5. $\dfrac{4}{9} = \dfrac{x}{72}$

5._____

6. $\dfrac{24}{18} = \dfrac{8}{x}$

6._____

7. Write a proportion for the sentence t is to 5 as 60 is to 25. Then solve.

7._____

8. Use a calculator to solve $\dfrac{12}{17} = \dfrac{x}{20}$. Round the result to 2 decimal places.

8._____

9. Find the lengths of the missing sides.

9._____

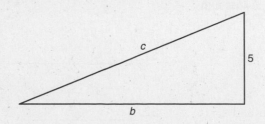

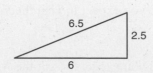

8.4 Short Quiz

Name_____

Date_____

In Exercises 1 and 2, use the following information: Tickets for the circus sell at the rate of 1560 per hour.

1. How many tickets are sold per minute?

1._____

2. After $6\frac{3}{4}$ hours, tickets are sold out. Every seat has been filled. What is the seating capacity of the venue (arena)?

2._____

3. If you pay $6.00 tax on a $119.95 item, about how much tax can you expect to pay on a $200.00 item?

3._____

In Exercises 4–7, solve the percent equation. Round your answer to 2 decimal places.

4. 65 is what percent of 500?

4._____

5. What is 12% of 24?

5._____

6. 18 is 20% of what number?

6._____

7. What is 250% of 200?

7._____

8. You predicted the outcome of 20 coin tosses with 40% accuracy. How many times was your prediction correct?

8._____

9. Imagine that you toss a coin 50 times. How many times do you think that the coin would land on heads? What percent of the total number of tosses would this be?

9.

In Exercises 1–3, tell whether the quotient is a rate or a ratio. Then simplify. (8.1)

1. 144 wheels/8 trucks

             1. _____

2. 3 classes/87 students

             2. _____

3. 60 cm/65 cm

             3. _____

4. Write three different proportions involving the sides of the triangles. (8.2)

             4.

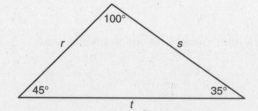

In Exercises 5 and 6, decide whether the equation is true. (8.2)

5. $\dfrac{3}{13} = \dfrac{12}{52}$

             5._____

6. $\dfrac{5}{16} = \dfrac{6}{17}$

             6._____

7. Zio's sells more pepperoni pizzas than plain pizzas by a ratio of 7 to 4. If 585 pepperoni pizzas are sold in a month, about how many plain pizzas are sold in the same time period? (8.3)

             7._____

In Exercises 8 and 9, solve the percent equation. (8.4)

8. What is 45% of 160?

             8._____

9. 42 is what percent of 300?

             9._____

In Exercises 10 and 11, solve the proportion. (8.2)

10. $\dfrac{x}{7} = \dfrac{12}{42}$

             10._____

11. $\dfrac{42}{21} = \dfrac{10}{z}$

             11._____

12. A high school has 700 students in its graduating class. Write the percent of students that you think might attend college. Then find the number of students that represent that percent. (8.4)

             12._____

In Exercises 1–3, tell whether the quotient is a rate or a ratio. Then simplify. (8.1)

1. 27 appointments/3 days

2. 18 inches/72 inches

3. 15 tanks/180 gallons of gas

4. Write three different proportions involving the sides of the triangles. (8.2)

1. _____

2. _____

3. _____

4. _____

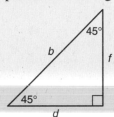

In Exercises 5 and 6, decide whether the equation is true. (8.2)

5. $\dfrac{4}{15} = \dfrac{15}{4}$

6. $\dfrac{8}{19} = \dfrac{32}{76}$

5. _____

6. _____

7. A jewelry-maker designed bracelets containing 3 large beads and 8 small beads. If she uses 600 small beads, how many large beads will she need? (8.3)

7. _____

In Exercises 8 and 9, solve the percent equation. (8.4)

8. What is 60% of 125?

9. 37 is what percent of 148?

8. _____

9. _____

In Exercises 10–12, solve the proportion. (8.2)

10. $\dfrac{16}{14} = \dfrac{a}{42}$

11. $\dfrac{4}{11} = \dfrac{b}{20}$

12. Roland attempted 24 shots in last night's basketball game. How many of the 24 shots do you think Roland made? What percent of the total number of shots would this be?

10. _____

11. _____

12. _____

Name _____

Date _____

In Exercises 1–3, use the following.

You take a survey in your school about favorite sports. The circle graph shows the percent of students who favored each sport. 75 students said that swimming was their favorite.

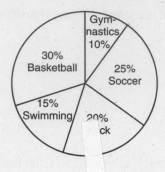

Gym-nastics 10%

30% Basketball

25% Soccer

15% Swimming

20% ck

1. How many students were surveyed?

1. _____

2. How many students selected soccer as their favorite?

2. _____

3. How many students chose track as their favorite?

3. _____

4. The new middle school will enroll 2250 students. Of these udents, 38% will be in grade 9. How many students is this?

4. _____

In Exercises 5 and 6, decide whether the change is an increase or a decrease. Then find the percent.

5. Before: 20, after: 25

5. _____

6. Opening price: $12.50, closing price: $9.50

6. _____

7. Describe a real-life situation in which you might find a 25% decrease. Then write a problem for the situation and solve it.

7.

8. Wayne scored 950 on a standardized test one year. After enrolling in a study course for the test, he took the test again and his score increased by 16%. What did Wayne score on the test the second time around?

8. _____

Name_____

Date _____

1. A fast food restaurant has 3 sandwich choices and 5 side dish choices. Use the Counting Principle to determine how many different combinations of 1 sandwich and 1 side dish can be made.

1._____

In Exercises 2 and 3, you are buying a watch. You can select a digital, standard, or Roman Numeral face, and a metal, leather, canvas, or plastic band.

2. How many different watches can you choose from?

2._____

3. Verify your answer with a tree diagram.

3._____

4. Use Pascal's Triangle to find the number of different ways to choose a pizza with 4 toppings if there are 7 topping choices.

4._____

In Exercises 5 and 6, use the following information.

Each of the spinners is spun once. On the first spinner, the arrow is equally likely to land on any number and on the second spinner, the arrow is equally likely to land on any letter.

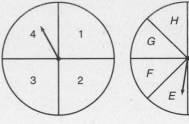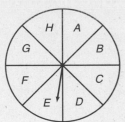

5. How many outcomes are possible?

5._____

6. List a possible outcome for which the probability is 1/16.

6._____

In Exercises 7 and 8, two number cubes are rolled simultaneously. Find the probability of each outcome.

7. A roll of a two and a three

7._____

8. A roll of an even number and a six

8._____

In Exercises 1 and 2, decide whether the quotient is a rate or a ratio. Then simplify. (8.1)

1. 15 kilometers/0.75 kilometers

1._____

2. 96 campers/8 bunks

2._____

In Exercises 3 and 4, write the verbal phrase as a rate or a ratio. State whether the result is a rate or a ratio. Then simplify. (8.1)

3. 46 out of 50 problems correct

3._____

4. 2 buses for 112 students

4._____

In Exercises 5–8, solve the proportion. (8.2)

5. $\dfrac{c}{5} = \dfrac{7}{12}$

5._____

6. $\dfrac{8}{15} = \dfrac{24}{d}$

6._____

7. $\dfrac{15}{2} = \dfrac{y}{4}$

7._____

8. $\dfrac{9}{34} = \dfrac{12}{v}$

8._____

In Exercises 9–12, solve the percent equation. (8.4)

9. 15 is what percent of 90?

9._____

10. 83 is 114% of what number?

10._____

11. What is 45 percent of 150?

11._____

12. What is 225 percent of 12?

12._____

In Exercises 13–16, use the similar
triangles. (8.2)

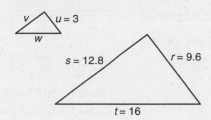

13. Find the ratio of *r* to *u*.

13._____

14. Solve for *v*.

14._____

15. Solve for *w*.

15._____

16. Find the ratio of *t* to *w*.

16._____

In Exercises 17–19, decide whether the change in the quantities represents
a percent increase or a percent decrease. Then find the percent. (8.6)

17. Before: 35, after: 50

17._____

18. Overcoat price in September: $100; overcoat price in April: $38

18._____

19. Write a situation that reflects a percent increase or a percent
decrease. Tell whether it is an increase or a decrease. Then find
the percent. (8.6)

19.

In Exercises 20 and 21, consider the following: Your family is purchasing a
new car. You may buy a van, a truck, or a station wagon. Your parents prefer
white, blue, red, black, and gold. (8.7)

20. List the different combinations of type and color.

20._____

21. Use the Counting Principle to confirm the number of combinations
in your list.

21._____

In Exercises 22–25, use the following information.

You push a button on the telephone without looking at it. Find the probability of each outcome. (8.7)

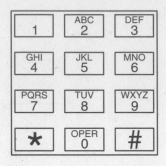

22. You touch a vowel.

22._____

23. You touch a button with no letters.

23._____

24. You touch a button whose letters can be rearranged to make a word.

24._____

25. You touch a prime number.

25._____

In Exercises 26–28, use the following information.

A telephone pole is casting a 50-foot shadow at the same time that a 12-foot tree is casting a 20-foot shadow. (8.3)

26. Draw a diagram that shows the telephone pole, the tree, and the two shadows.

26.

27. Write a proportion that involves the height of the telephone pole, the height of the small tree, and the lengths of the two shadows.

27._____

28. Solve the proportion to find the height of the pole.

28._____

In Exercises 29 and 30, Albee's Shoe Store is having a "Going Out of Business" Sale. All shoes are 40% off and all socks are 25% off. (8.5)

29. How much will a pair of shoes cost that were regularly $59.95?

29._____

30. How much will a 6-pack of socks be that were regularly $12.50?

30._____

In Exercises 1–4, decide whether the quotient or verbal phrase is a rate or a ratio. Find its simplified form. (8.1)

1. 162 seats/9 rows

   a. rate: $\dfrac{162 \text{ seats}}{9 \text{ rows}}$ **b.** rate: $\dfrac{18 \text{ seats}}{1 \text{ row}}$ **c.** ratio: $\dfrac{162}{9}$ **d.** ratio: $\dfrac{18}{1}$

   1._____

2. 3.5 liters/1.5 liters

   a. rate: $\dfrac{3.5}{1.5}$ **b.** rate: $\dfrac{7 \text{ liters}}{3 \text{ liters}}$ **c.** ratio: $\dfrac{3.5}{1.5}$ **d.** ratio: $\dfrac{7}{3}$

   2._____

3. 14 rainy days out of 30 days

   a. ratio: $\dfrac{14}{30}$ **b.** ratio: $\dfrac{7}{15}$ **c.** rate: $\dfrac{14 \text{ days}}{30 \text{ days}}$ **d.** rate: $\dfrac{7}{15}$

   3._____

4. 315 minutes/7 classes

   a. rate: $\dfrac{315 \text{ minutes}}{7 \text{ classes}}$ **b.** ratio: $\dfrac{45}{1}$

   c. ratio: $\dfrac{315}{7}$ **d.** rate: $\dfrac{45 \text{ minutes}}{1 \text{ class}}$

   4._____

In Exercises 5–8, solve the proportion. (8.2)

5. $\dfrac{r}{27} = \dfrac{8}{18}$

   a. $r = 9$ **b.** $r = 16$ **c.** $r = 12$ **d.** $r = 36$

   5._____

6. $\dfrac{12}{30} = \dfrac{10}{n}$

   a. $n = 25$ **b.** $n = 28$ **c.** $n = 4$ **d.** $n = 8$

   6._____

7. $\dfrac{10}{m} = \dfrac{9}{5}$

   a. $m = 5$ **b.** $m = 6$ **c.** $m = \dfrac{5}{9}$ **d.** $m = \dfrac{50}{9}$

   7._____

8. $\dfrac{7}{x} = \dfrac{4}{9}$

   a. $x = \dfrac{63}{4}$ **b.** $x = 6\dfrac{3}{4}$ **c.** $x = 12$ **d.** $x = \dfrac{9}{4}$

   8._____

9. Hal's trail mix contains peanuts to raisins in a ratio of 8 to 5. If Hal uses 325 g of raisins, how many grams of peanuts must he use? (8.3)

    a. 600   **b.** 500   **c.** 475   **d.** 520

9._____

In Exercises 10–13, solve the percent equation. (8.4)

10. What is 90 percent of 50?

    a. 90   **b.** 45   **c.** 5   **d.** 10

10._____

11. 30 is what percent of 20?

    a. 150%   **b.** 66.66%   **c.** 60%   **d.** $1\frac{1}{2}\%$

11._____

12. 14 is 35% of what number?

    a. 20   **b.** 40   **c.** 30   **d.** 39

12._____

13. What is 18% of 54?

    a. 9   **b.** 10.8   **c.** 9.72   **d.** 10

13._____

In Exercises 14 and 15, find the percent increase or decrease that the change represents. (8.6)

14. Before: 26, after: 40

    a. increase, 14%   **b.** increase, 35%

    c. increase, ≈ 53.8%   **d.** decrease, ≈ 53.8%

14._____

15. Before: 310, after: 247

    a. increase, ≈ 20.32%   **b.** increase, ≈ 19.98%

    c. decrease, ≈ 25.5%   **d.** decrease, ≈ 20.32%

15._____

In Exercises 16 and 17, find the percent increase or decrease that the quantities represent. (8.6)

16. Last year's cost: $420, this year's cost: $375

    a. decrease, ≈ 10.7%   **b.** decrease, ≈ 14.66%

    c. increase, ≈ 14.66%   **d.** increase, ≈ 10.7%

16._____

17. June pool attendance: 247, August pool attendance: 173

    a. decrease, ≈ 29.96%   **b.** increase, ≈ 29.96%

    c. decrease, ≈ 42.77%   **d.** increase, ≈ 42.77%

17._____

18. A gold chain cost $125 ten years ago. In the last 10 years, the price has increased 420%. What is the price of the gold chain today? (8.5)

   a. $575   **b.** $650   **c.** $620   **d.** $625

18._____

19. How many possible ways can you order a hamburger with a choice of two types rolls and 7 toppings? (8.7)

   a. 49   **b.** 98   **c.** 14   **d.** 9

19._____

20. How many ways can you choose 3 CD's from a selection of 5 CD's? (8.7)

   a. 15   **b.** 8   **c.** 13   **d.** 10

20._____

In Exercises 21–23, use the triangles. (8.2)

21. What is the ratio of w to f?

   a. 6 to 5   **b.** 3 to 2

   c. 10 to 9   **d.** 2 to 1

21._____

22. What is the measure of side x?

   a. 12 cm   **b.** 8 cm   **c.** 9 cm   **d.** 10 cm

22._____

23. What is the measure of side y?

   a. 10.2 cm   **b.** 20 cm   **c.** 10 cm   **d.** 12 cm

23._____

In Exercises 24–27, use the spinner and the number cube. (8.7)

24. What is the probability of getting a total of 12?

   a. $\frac{1}{6}$ **b.** $\frac{1}{12}$

   c. $\frac{1}{36}$ **d.** $\frac{1}{2}$

24._____

25. What is the probability of getting a total of 8?

   a. $\frac{1}{6}$ **b.** $\frac{2}{9}$ **c.** $\frac{5}{36}$ **d.** $\frac{3}{12}$

25._____

26. What is the probability of getting a total that is less than 4?

   a. $\frac{1}{12}$ **b.** $\frac{1}{9}$ **c.** $\frac{1}{6}$ **d.** $\frac{1}{10}$

26._____

27. What is the probability that you will get an odd number on both the spinner and the number cube?

   a. $\frac{1}{12}$ **b.** $\frac{1}{18}$ **c.** $\frac{1}{6}$ **d.** $\frac{1}{4}$

27._____

In Exercises 1 and 2, decide whether the quotient is a rate or a ratio. Then simplify. (8.1)

1. 121 players/11 teams

   1._____

2. 86 degrees/68 degrees

   2._____

3. Write a verbal phrase that could be expressed as a rate. Then write the rate and simplify it. (8.1)

   3._____

4. Write a verbal phrase that could be expressed as a ratio. Then write the ratio and simplify it. (8.1)

   4._____

In Exercises 5–7, solve the proportion. (8.2)

5. $\dfrac{15}{7} = \dfrac{5}{y}$

   5._____

6. $\dfrac{z}{12} = \dfrac{4}{7}$

   6._____

7. $\dfrac{2}{3} = \dfrac{(a + 1)}{12}$

   7._____

8. Write a proportion for which the value of the variable is 3. (8.2)

   8._____

In Exercises 9–11, solve the percent equation. (8.4)

9. What percent of 25 is 7?

   9._____

10. 25 is 12.5% of what number?

    10._____

11. What is 325% of 13?

    11._____

In Exercises 12–14, use the similar triangles. (8.2)

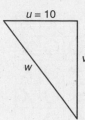

$u = 10$

$x = 6$

v

w

$z = 10$ $y = 8$

12. Find the ratio of u to x.

13. Find the value of v.

14. Find the value of w.

12._____

13._____

14._____

In Exercises 15 and 16, decide whether the quantities represent a percent increase or a percent decrease. Then find the percent. (8.6)

15. Cost before tax: $15.00, cost after tax: $15.90

15._____

16. Enrollment in September: 765, enrollment in January: 790

16._____

17. Describe how you could estimate 15% of $50 using mental math. Then solve. (8.4)

17.

18. Thirty-five percent of Kim's karate class is children under the age of 12. How many of her 180 students are 12 or older? (8.5)

18._____

In Exercises 19 and 20, use the following information.

In Los Angeles, California, the 750-foot Two California Plaza casts a 37.5 foot shadow at the same time that the AT&T Building casts a 31-foot shadow. (8.3)

(Source: 1997 World Almanac)

19. Write a proportion that involves the height of the two buildings and the lengths of the two shadows.

19._____

20. Solve the proportion to find the height of the AT&T Building.

20._____

In Exercises 21–23, use the square game board. Each small square is the same size. (8.4, 8.7)

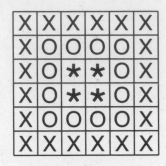

21. What percent of the game board has squares containing an X?

21._____

22. What percent of the game board has squares containing an O?

22._____

23. What is the probability that a coin dropped on the game board would land on a square containing a *?

23._____

24._____

24. You are purchasing a telephone. You may choose a wall, desk, or antique style. You may select a touch-tone or rotary dial. Color choices are white, ivory, tan, and black. How many possible choices of telephones are there? (8.7)

25. How many ways can you choose 2 candidates from a 7-candidate field? (8.7)

25._____

In Exercises 26–28, one card is drawn from a deck of 52 playing cards. (8.7)

26. What is the probability of drawing a red card?

26._____

27. What is the probability of drawing an ace?

27._____

28. Do you have a better chance of drawing a face card than of drawing a heart? Explain.

28.

29. You take a job in an automotive shop for $4.75 an hour. One year later, your boss offers you the choice of an 8% raise or an increase of $0.30 an hour. Which should you take? Explain. (8.5)

29.

Name_____

Date _____

In Exercises 1 and 2, write both square roots of the number.

1. 81

1._____

2. $\dfrac{9}{4}$

2._____

In Exercises 3 and 4, write both solutions of the equation.

3. $y^2 + 7 = 56$

3._____

4. $2x^2 = 128$

4._____

5. Write an algebraic equation for the sentence, t squared decreased by 1 is 24. Then solve.

5._____

6. The small squares are each 1 square unit. Estimate the side lengths of the shaded square.

6._____

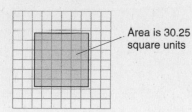

Area is 30.25 square units

7. Write a number that is irrational. Explain why it is irrational.

7._____

In Exercises 8 and 9, evaluate the expression for $x = 3$, $y = 9$, and $z = 25$. Tell whether the result is rational.

8. $\sqrt{y} + \sqrt{x}$

8._____

9. $\sqrt{y} \div \sqrt{z}$

9._____

In Exercises 1–3, *a* and *b* are the lengths of the legs of a right triangle, and *c* is the length of the hypotenuse. Find the missing length. Round your results to three decimal places, if necessary.

1. $a = 4, b = 8$

2. $a = 5, c = 13$

3. $b = 9, c = 20$

4. Find the length of the third side of the triangle below. Round your results to two decimal places.

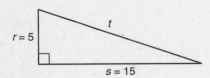

5. Find the perimeter and area of the figure.

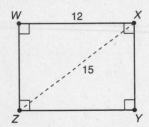

6. How high was the tree before it was hit by lightning?

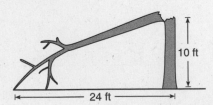

7. Draw a triangle similar to the triangle below. Write the measurements of each side.

1. _____

2. _____

3. _____

4. _____

5.

6. _____

7.

Mid-Chapter **9** Test | Form A

(Use after Lesson 9.4)

Name_____

Date_____

In Exercises 1 and 2, find both square roots of the number. (9.1)

1. 169

1._____

2. 0.04

2._____

3. Find the side lengths of the square. Round your result to three decimal places. (9.1)

Area is 42.3 square units

3._____

In Excrcises 4 and 5, decide whether the number is rational or irrational. (9.2)

4. $\sqrt{45}$

4._____

5. $\sqrt{1.44}$

5._____

In Exercises 6 and 7, solve the triangle. Round your result to three decimal places. (9.3)

6.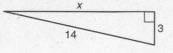

6._____

7.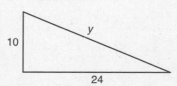

7._____

8. A chimney sweep must place a 52-foot ladder against the wall of a home so that it reaches a point exactly 48 feet high. How far from the home should he place the ladder? (9.4)

8._____

In Exercises 1 and 2, find both square roots of the number. (9.1)

1. 256

2. 0.09

3. Find the side lengths of the square. Round your result to three decimal places. (9.1)

Area is 47.5 square units

1. _____

2. _____

3. _____

In Exercises 4 and 5, decide whether the number is rational or irrational. (9.2)

4. $\sqrt{50}$

5. $\sqrt{2.25}$

4. _____

5. _____

In Exercises 6 and 7, solve the triangle. Round your result to three decimal places. (9.3)

6.
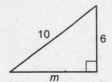

7.
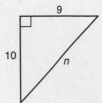

8. A wooden plank placed 50 cm from an apartment building touches an apartment window 1.2 m (120 cm) from the ground. What is the length of the plank? (9.4)

6. _____

7. _____

8. _____

Name_____

Date _____

In Exercises 1 and 2, graph the inequality.

1. $x > 3$

2. $x \leq -4$

In Exercises 3 and 4, write the inequality represented by the graph.

3.

4.

5. Write an algebraic model for the phrase, w increased by 2 is less than 14. Then solve.

In Exercises 6–9, solve the inequality. Then graph the solution.

6. $y + 5 > 3$

7. $n - 4 \leq 0$

8. $8m > 2$

9. $-\dfrac{4}{5} \geq -\dfrac{9}{10}x$

10. You are making awards for a track and field event. You need to buy $\frac{5}{8}$ yard of ribbon for each award. You must make at least 150 awards. If you can't buy part of a yard, how many yards of ribbon must you buy?

1. _____

2. _____

3. _____

4. _____

5. _____

6. _____

7. _____

8. _____

9. _____

10. _____

9.8 Short Quiz

Name_____

Date _____

In Exercises 1–3, solve the inequality.

1. $y < 3y - 2$

1._____

2. $\frac{1}{3}m + 1 \geq 4$

2._____

3. $5g - 8 > 3(g + 1)$

3._____

In Exercises 4 and 5, let n, $n + 1$, and $n + 2$ be consecutive integers. Write the inequality that represents the verbal sentence. Then solve the inequality.

4. The sum of 2 consecutive integers is less than or equal to -21.

4._____

5. The sum of 3 consecutive integers is greater than 61.

5._____

6. Can the side lengths for the triangle be correct? Explain.

6.

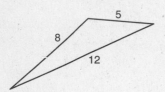

7. Can the numbers 0.325, 0.525, 0.875 be side lengths of a triangle? Explain.

7.

8. Draw a triangle with labeled side lengths $\frac{3}{10}$ and $\frac{5}{10}$. Write one value that could not be a third side length, and one value that could be a third side length. Explain.

8.

In Exercises 1–3, list both square roots. (9.1)

1. 625

2. 1.96

3. $\dfrac{49}{81}$

1._____

2._____

3._____

In Exercises 4–6, solve the equation. (9.1)

4. $90 = d^2 + 9$

5. $6z^2 = 36$

6. $100 = f^2 - 21$

4._____

5._____

6._____

In Exercises 7–10, match each number with its location on the number line. (9.2)

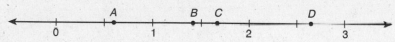

7. $\dfrac{5}{3}$

8. $\dfrac{3}{5}$

9. $\sqrt{2}$

10. $\sqrt{7}$

7._____

8._____

9._____

10._____

11. Choose an irrational number between 0 and 3 and add it to the number line. (9.2)

11._____

12. A square has an area of 12.34 square units. Find the length of each side. Round your answer to 3 decimal places. (9.1)

12._____

13. A right triangle has sides of 6 feet and 10 feet. Name two possibilities for the length of the third side. (9.3)

13._____

In Exercises 14–16, solve the right triangle. Round results to three decimal places, if necessary. (9.3)

14.

14._____

15.

15._____

16.

16._____

In Exercises 17–20, graph the inequality on a number line. (9.5)

17. $x \leq 0$

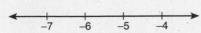

17.

18. $x > 8$

18.

19. $x < -5$

19.

20. $x \geq -2$

20.

Passport to Algebra and Geometry

In Exercises 21–25, solve the inequality. (9.6, 9.7)

21. $k + 15 < 7$

21._____

22. $-5 > -y - 6$

22._____

23. $\frac{3}{5}t \leq \frac{2}{5}$

23._____

24. $-5g \geq -28$

24._____

25. $2(-3 - w) < 5w - 12$

25._____

In Exercises 26 and 27, use the triangle. (9.8)

26. What is the range for the measure of the third side?

26._____

27. The perimeter of the triangle must be between what two numbers?

27._____

In Exercises 28 and 29, use the following information.

A kitten is trapped at the top of a tree. You place a 26-foot ladder 10 feet from the tree. It places you right at the top of the tree where you can rescue the kitten. (9.4)

28. Draw a sketch showing the ladder and the tree.

28.

29. How tall is the tree?

29._____

In Exercises 1–3, what are the square roots for the number? (9.1)

1. 324

   a. $+18$ and $+36$

   c. $3\sqrt{100} + \sqrt{24}$

   b. $+18$ and -18

   d. $\sqrt{+324}$ and $\sqrt{-324}$

1._____

2. 1.21

   a. $+1.1$ and -1.1

   c. $+11$ and -11

   b. $1 + \sqrt{0.21}$

   d. $\sqrt{+1.21}$ and $\sqrt{-1.21}$

2._____

3. $\dfrac{25}{100}$

   a. $+\dfrac{1}{4}$ and $-\dfrac{1}{4}$

   c. $+\dfrac{1}{8}$ and $-\dfrac{1}{8}$

   b. $\dfrac{\sqrt{25}}{100}$

   d. $+\dfrac{1}{2}$ and $-\dfrac{1}{2}$

3._____

In Exercises 4 and 5, solve the equation. (9.1)

4. $76 = f^2 - 5$

   a. $f = +9$

   c. $f = +8.426$

   b. $f = +9$ and -9

   d. $f = -\sqrt{81}$

4._____

5. $5b^2 = 8.45$

   a. $b = +1.3$ and -1.3

   c. $b = +1.857$

   b. $b = +1.69$ and -1.69

   d. $b = 1.69$

5._____

6. If a square has an area of 89.7 square units, what is the length of each side?

   a. ≈ 9.47 units **b.** ≈ 29.94 units **c.** ≈ 2.99 units **d.** ≈ 22.43 units

6._____

In Exercises 7 and 8, solve the right triangle. (9.3)

7.

   a. $x = 14$

   b. $x = 18$

   c. $x = 22$

   d. $x = 20$

7._____

8.

   a. $x = 18$

   b. $x = 16.5$

   c. $x = 21.5$

   d. $x = 20$

8._____

Passport to Algebra and Geometry

In Exercises 9–12, what letter corresponds to the number on the number line? (9.2)

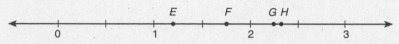

9. $\sqrt{5}$ 9._____

    a. E   **b.** F   **c.** G   **d.** H

10. $\frac{7}{3}$ 10._____

    a. E   **b.** F   **c.** G   **d.** H

11. $\sqrt{1.44}$ 11._____

    a. E   **b.** F   **c.** G   **d.** H

12. $\frac{7}{4}$ 12._____

    a. E   **b.** F   **c.** G   **d.** H

13. Which point would NOT appear on the portion of the line shown above? (9.2)   13._____

    a. $\sqrt{2.56}$ **b.** $\frac{8}{3}$ **c.** $\sqrt{17}$ **d.** 1.99

In Exercises 14–16, which inequality corresponds to the graph? (9.5)

14.   14._____

    a. $x \leq 2$ **b.** $x \geq 2$ **c.** $x > 2$ **d.** $x < 2$

15.   15._____

    a. $x < -7$ **b.** $x > -7$ **c.** $x \geq -7$ **d.** $x \leq -7$

16.   16._____

    a. $x \geq \frac{5}{3}$ **b.** $x \leq \frac{5}{3}$ **c.** $x < \frac{5}{3}$ **d.** $x > \frac{5}{3}$

17. Which graph shows the inequality $x \geq -1.3$? (9.5)   17._____

a.
-1.4 -1.3 -1.2 -1.1

b.
-1.4 -1.3 -1.2 -1.1

c.
-1.4 -1.3 -1.2 -1.1

d.
-1.4 -1.3 -1.2 -1.1

18. A triangle has two sides that measure 9 feet and 15 feet. Which   18._____
could be a possible value for the third side of the triangle? (9.8)

  a. 12   **b.** 25   **c.** 6   **d.** -6

In Exercises 19–22, solve the inequality. (9.6, 9.7)

19. $-4 - y > 17$ 19._____

  a. $y > 21$ **b.** $y > -21$ **c.** $y < -21$ **d.** $y > -21$

20. $6g < -45$ 20._____

  a. $g < 7.5$ **b.** $g > 7.5$ **c.** $g < -7.5$ **d.** $g > -7.5$

21. $\frac{4}{7}h \leq -\frac{8}{15}$ 21._____

  a. $h \leq -\frac{32}{105}$ **b.** $h \geq -\frac{14}{15}$ **c.** $h \leq \frac{14}{15}$ **d.** $h \leq -\frac{14}{15}$

22. $3(e - 4) > 4e - 15$ 22._____

  a. $e > 3$ **b.** $e < -3$ **c.** $e < 3$ **d.** $e > -3$

In Exercises 23 and 24, use the following verbal model:
The product of 7 and $(3 - x)$ is greater than 50. (9.7)

23. Write the algebraic model for the sentence.   23._____

  a. $7 + (3 - x) > 50$ **b.** $7(3 - x) > 50$

  c. $7(3 - x) < 50$ **d.** $7 + (3 - x) \geq 50$

24. For what value of x is the inequality true?   24._____

  a. $x > -40$ **b.** $x > -\frac{29}{7}$ **c.** $x < -\frac{29}{7}$ **d.** $x < -40$

25. A ladder placed 12.5 feet from a building touches a building exactly   25._____
30 feet from the ground. What is the length of the ladder? (9.4)

  a. 32.5 feet   **b.** 35 feet   **c.** 17.5 feet   **d.** 9.219 feet

In Exercises 1–3, list both square roots. (9.1)

1. 529

1. _____

2. 0.0036

2. _____

3. $\dfrac{121}{225}$

3. _____

In Exercises 4–6, solve the equation. (9.1)

4. $147 + x^2 = 228$

4. _____

5. $\frac{1}{8}x^2 - 12 = 6$

5. _____

6. $16 = 4x^2 - 6$

6. _____

In Exercises 7–9, match each number with its location on the number line. (9.2)

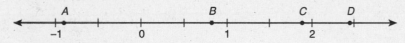

7. $\sqrt{6}$

7. _____

8. $-\sqrt{0.81}$

8. _____

9. $\dfrac{14}{17}$

9. _____

10. Add a point to the number line whose square root value is between 1 and 2. Label it point E. Write the number that describes its location. (9.2)

10. _____

11. Add point F to the number line. Make its value < 0. Write the number that describes its location. (9.2)

11. _____

12. A square has an area of 42.37 square inches. Find the length of each side. Round your answer to three decimal places. (9.1)

12. _____

13. A right triangle has two side lengths of 8.25 inches and $\sqrt{77.44}$ inches. List two possibilities for the length of the third side. Round your answers to three decimal places. (9.8)

13. _____

In Exercises 14–16, solve the right triangle. Round your answers to two decimal places. (9.3)

14.

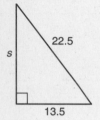

14. _____

15.

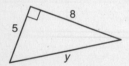

15. _____

16.

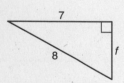

16. _____

17. One side length of a right triangle is 7 units. Use the Pythagorean Theorem to find possible lengths for the other two sides. Then draw and label the triangle. (9.3)

17. _____

In Exercises 18–20, solve the inequality. Then graph the solution. (9.5–9.7)

18. $8x - 7 < 14$

18. _____

19. $\frac{3}{4}t - 2 \leq \frac{5}{6}t + 5$

19. _____

20. $5(4.2 - b) > 18.5$

20. _____

In Exercises 21–23, write an algebraic model for the sentence. Then solve. (9.7)

21. Three times *s* increased by seven is less than ten.

21._____

22. Fourteen and six tenths is less than or equal to negative three times *q*.

22._____

23. The product of -5 and $(x - 7)$ is less than -15.

23._____

In Exercises 24 and 25, use the following information.

You have $5.00 to spend at an arcade. You spend $0.95 on a drink and $1.25 on pretzels. You can spend the rest of your money on video games at $0.65 each or on pinball games, priced at 2 games for $0.75 or $0.50 each. (9.7)

24. What is the maximum number of video games that you can play?

24._____

25. What is the maximum number of pinball games that you can play?

25._____

In Exercises 26–28, use the following information.

You are cleaning windows in your home. The first floor windows are 15 feet from the ground, and the second floor windows are 25 feet from the ground. You use a 30-foot ladder to reach both windows. (9.4)

26. How far away from the house is the ladder if its top rung reaches the first story windows?

26._____

27. How far away from the house is the ladder if its top rung reaches the second story windows?

27._____

28. Draw a diagram showing the ladder leaning against one set of windows. Label all the sides of the triangle formed by the ladder, the house, and the ground.

28._____

29. A door is 78 inches high and 36 inches wide. Can a thin piece of wood that is 7 feet wide be carried through the doorway? Explain your answer. (9.4).

29.

In Exercises 1–6, simplify the expression. (7.1, 7.2)

1. $\frac{3}{10} + \frac{7}{10}$

2. $\frac{11}{15} - \frac{8}{15}$

3. $\frac{4y}{9} + \frac{2y}{9}$

4. $\frac{2}{15} + \frac{1}{10}$

5. $\frac{3}{4} + \frac{4}{5}$

6. $\frac{2}{3}w - \frac{5}{8}w$

In Exercises 7–10, use a calculator to evaluate the expression. Round the result to three decimal places. (7.3)

7. $\frac{23}{35} + \frac{42}{51}$

8. $\frac{8}{9} - \frac{12}{25}$

9. $2 - \left(\frac{6}{7} + \frac{10}{11}\right)$

10. $\frac{23}{4}g + \frac{42}{124}g$

In Exercises 11 and 12, find the perimeter and area of the figure. (7.1–7.5)

11.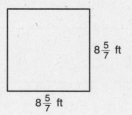

$8\frac{5}{7}$ ft

$8\frac{5}{7}$ ft

12.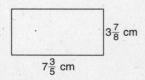

$3\frac{7}{8}$ cm

$7\frac{3}{5}$ cm

1. _____

2. _____

3. _____

4. _____

5. _____

6. _____

7. _____

8. _____

9. _____

10. _____

11. _____

12. _____

In Exercises 13–16, simplify the expression. (7.4, 7.5)

13. $\frac{6}{11} \cdot \frac{33}{44}$

13. _____

14. $\frac{3}{8} \cdot \frac{5}{6}$

14. _____

15. $24 \cdot \left(\frac{3}{8}\right)r$

15. _____

16. $\frac{7t}{5} \div \frac{14}{15}$

16. _____

In Exercises 17–20, write the portion as a decimal and a percent. (7.6, 7.7)

17. $\frac{1}{4}$

17. _____

18. $\frac{12}{25}$

18. _____

19. $\frac{115}{150}$

19. _____

20. $\frac{355}{500}$

20. _____

21. What portion of the square is shaded? Express your answer as a fraction, a decimal, and a percent. (7.6, 7.7)

21. _____

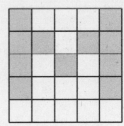

22. You borrow $525 from a bank that charges 8.5% annual interest. After one year, you repay the loan. How much do you owe? (7.8, 7.9)

22. _____

23. You pay 5% tax on a $95.00 purchase. What is your total cost? (7.8, 7.9)

23. _____

In Exercises 24–26, find the percent of the number. (7.8)

24. 35% of 80

24. _____

25. 225% of 36

25. _____

26. 10.5% of 250

26. _____

27. Your scores on five quizzes are 98%, 93%, 88%, 92%, and 86%. What is your average percent? (7.9)

27. _____

In Exercises 28–31, decide whether the quotient is a rate or a ratio. Then simplify. (8.1)

28. 12 swimmers/2 heats

28. _____

29. 8.6 miles/2 miles

29. _____

30. 12 months/52 weeks

30. _____

31. 126 cm/144 cm

31. _____

In Exercises 32–34, use the following information. (8.2, 8.3)

Television shows are evaluated using Nielson ratings. Nielson ratings help networks determine a show's popularity and how much to charge for advertising a product or service on that show. One Nielson ratings point represents 942,000 TV households.

32. During one week of Nielson ratings, the top show received a rating of 16.7. Approximately how many TV households were tuned in to the top-rated show?

32. _____

33. The tenth-rated show had approximately 11,635,800 TV sets tuned in during the ratings evaluation. How many Nielson points did it receive?

33. _____

34. One advertising agency purchases time only on shows that have a minimum of 10 million viewers. What is the lowest acceptable Nielson rating for this agency?

34. _____

In Exercises 35–38, solve the proportion. (8.2)

35. $\dfrac{2}{15} = \dfrac{x}{240}$

35. _____

36. $\dfrac{63}{144} = \dfrac{t}{16}$

36. _____

37. $\dfrac{7}{j} = \dfrac{4}{9}$

37. _____

38. $\dfrac{1.8}{y} = \dfrac{12.6}{15}$

38. _____

39. The triangles below are similar. Solve for x. (8.2)

39. _____

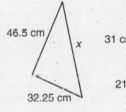

46.5 cm x 31 cm 38.5 cm

21.5 cm

32.25 cm

In Exercises 40–45, solve the percent equation. (8.4)

40. 8 is what percent of 40?

40. _____

41. What is 18% of 54?

41. _____

42. 7 is 12.5% of what number?

42. _____

43. 135 is what percent of 144?

43. _____

44. What is 4.5% of 60?

44. _____

45. 50 is 250 percent of what number?

45. _____

46. The Normans' total bill for dinner was $92.45. About how much money should they pay (altogether) if they want to leave about a 20% tip? (8.5)

46. _____

47. Explain how you might find 12% of 50 using mental math. Then solve. (8.4)

47. _____

In Exercises 48–54, use the following.

You take a survey to determine favorite electives among middle school students. The circle graph shows the result of your survey. Forty people said that their first choice was music. (8.5)

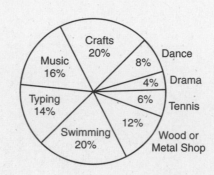

48. How many people were surveyed?

48. _____

49. How many people chose drama?

49. _____

50. How many people chose swimming or crafts?

50. _____

51. How many people chose shop?

51. _____

52. How many people chose typing?

52. _____

53. How many people chose tennis?

53. _____

54. How many people chose dance?

54. _____

In Exercises 55–57, decide whether the change is an increase or a decrease. Then find the percent. (8.6)

55. September: 19 words per minute; June: 41 words per minute

55. _____

56. Regular price: $29.99; Sale price: $23.99

56. _____

57. 1995–1996 Season: 32,576 season ticket holders
1996–1997 Season: 40,102 season ticket holders

57. _____

Passport to Algebra and Geometry

In Exercises 58–61, a number cube is rolled, then a red card is chosen from a deck (there are 26 red cards in a deck). Find the probability of each outcome. (8.7, 8.8)

58. A four and the five of hearts

58. _____

59. A three and a red two

59. _____

60. An even number and a heart

60. _____

61. A number less than 3 and the king of diamonds

61. _____

In Exercises 62–65, find the value of each number. Round your answer to two decimal places. Then graph the numbers on the number line. (9.1, 9.2)

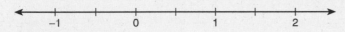

62. $A = -\sqrt{1.5}$

62. _____

63. $B = \frac{15}{16}$

63. _____

64. $C = -\sqrt{\dfrac{144}{169}}$

64. _____

65. $D = \sqrt{3}$

65. _____

66. Write a number whose square root is between 0 and 0.5. Graph it on the number line with the label E. (9.1, 9.2)

66. _____

In Exercises 67–69, solve the right triangle. Round your results to two decimal places. (9.3)

67.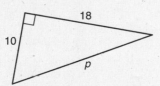

67. _____

68.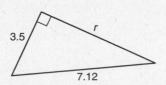

68. _____

69.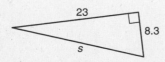

69. _____

In Exercises 70–74, solve the inequality. Then graph the solution. (9.4–9.7)

70. $b - 1.8 < 14.2$

70. _____

71. $3i - 15 \geq 34$

71. _____

72. $\frac{1}{3}f + \frac{3}{5} > \frac{2}{3}f$

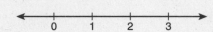

72. _____

73. $\frac{4}{9}(r - 3) \leq \frac{2}{3}$

73. _____

74. $\frac{3}{2}(4 - 6k) \leq 2$

74. _____

In Exercises 75–78, decide whether the triangle can have the given side lengths. Explain. (9.8)

75.

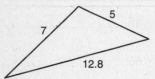

75. _____

76.

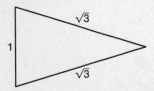

76. _____

77.

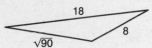

77. _____

78.

78. _____

Passport to Algebra and Geometry

In Exercises 79–81, write an algebraic model for the sentence. Then solve. (9.1, 9.6)

79. The sum of r^2 and 14 is 135.

79. _____

80. The product of 4 and y^2 is 64.

80. _____

81. t divided by -7 is greater than or equal to -5.

81. _____

82. A right triangle has two side measures of 7 units and 10 units. List two possibilities for the measure of the third side. Then sketch triangles using both of the possible measures. (9.3, 9.4)

82. _____

In Exercises 83–85, use the rectangle at the right. The perimeter is 48 inches. (9.3, 9.4)

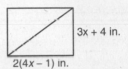
3x + 4 in.
2(4x − 1) in.

83. Write an equation that can be used to find the value of x. Then solve for x.

83. _____

84. Write the dimensions of the rectangle.

84. _____

85. What is the length of the diagonal line running through the rectangle?

85. _____

In Exercises 86–88, use the following information. (8.5)

A store is having a special sale. Items priced between $15 and $29.99 are 10% off, and items priced between $30 and $44.99 are 15% off.

86. What is the most you can save on an item that is 10% off?

86. _____

87. Which would cost more, an item originally priced at $25.99 or at $32.99? Explain.

87. _____

88. You purchase a shirt for $27.00 and a pair of shorts for $32.50. What is your total cost, excluding tax?

88. _____

Name _____

Date _____

In Exercises 1–8, use the diagram.

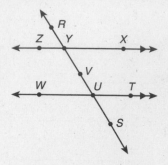

1. Write four other names for the line $\overleftrightarrow{UV}$.

1. _____

2. Name two different line segments that lie on $\overleftrightarrow{WT}$.

2. _____

3. Name four rays that have the same beginning point.

3. _____

4. Name two parallel lines.

4. _____

5. Name two pairs of intersecting lines.

5. _____

6. Name four acute angles.

6. _____

7. Name four obtuse angles.

7. _____

8. Name three straight angles.

8. _____

In Exercises 9 and 10, use a protractor to draw the angle.

9. Draw an acute angle. Find its measure.

9. _____

10. Draw an obtuse angle. Find its measure.

10. _____

10.4 Short Quiz

Name_____

Date _____

In Exercises 1–4, use the diagram.

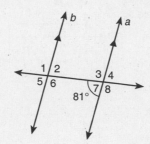

1. Which two lines are parallel?

2. List all angles whose measures are 81°.

3. Name two corresponding angles that have the same measure.

4. Name four pairs of vertical angles.

5. Explain why angles 1 and 2 are
   congruent in the diagram at the right.

1._____

2._____

3._____

4._____

5.

In Exercises 6 and 7, draw a figure that has the given characteristics.

6. Exactly one line of symmetry

7. Rotational and line symmetry

8. Trace the figure and find out what word
   is spelled using the indicated line of
   symmetry.

6._____

7._____

8._____

In Exercises 1–3, use the rectangular prism. (10.1)

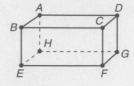

1. Name another point that lies in the same plane as *B, E,* and *C.*

1._____

2. Name two lines that are parallel to $\overleftrightarrow{DG}$.

2._____

3. Name the point of intersection of $\overline{CF}$ and $\overline{GF}$.

3._____

In Exercises 4–7, use the figure. (10.2)

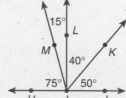

4. Name the right angles.

4._____

5. Name the obtuse angles.

5._____

6. Name the acute angles.

6._____

7. Name the straight angle.

7._____

In Exercises 8–10, draw the triangle. (10.5)

8. Scalene obtuse

8.

9. Isosceles acute

9.

10. Equilateral, acute, equiangular

10.

11. Identify any symmetry of the figure. (10.4)

11._____

In Exercises 12–14, use the figure. (10.3)

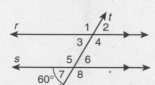

12. Which two lines are parallel?

12._____

13. Name four pairs of vertical angles.

13._____

14. List all angles whose measure is 120°.

14._____

Mid-Chapter 10 Test   Form B

(Use after Lesson 10.5)

Name _____

Date _____

In Exercises 1–3, use the rectangular prism. (10.1)

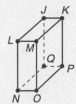

1. Name another point that lies in the same plane as *K*, *P*, and *M*.

2. Name two lines that are parallel to $\overleftrightarrow{JK}$.

3. Name the point of intersection of $\overline{LN}$ and $\overline{ON}$.

In Exercises 4–7, use the figure. (10.2)

4. Name the right angles.

5. Name the obtuse angles.

6. Name the acute angles.

7. Name the straight angle.

In Exercises 8–10, draw the triangle. (10.5)

8. Scalene acute

9. Isosceles right

10. Equilateral, acute, equiangular

11. Identify any symmetry of the figure. (10.4)

In Exercises 12–14, use the figure. (10.3)

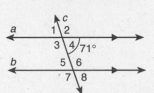

12. Which two lines are parallel?

13. Name four pairs of corresponding angles.

14. List all angles whose measure is 109°.

1. _____

2. _____

3. _____

4. _____

5. _____

6. _____

7. _____

8.

9.

10.

11. _____

12. _____

13. _____

14. _____

In Exercises 1–3, sketch the indicated type of triangle. Then label it with appropriate side or angle measures.

1. Obtuse isosceles

2. Right scalene

3. Equilateral

4. Draw $\triangle DEF$ with angle measures $m\angle D = 70°$, $m\angle E = 90°$, $m\angle F = 20°$. Then classify it according to its sides and angles.

5. Classify the triangle according to its sides and angles.

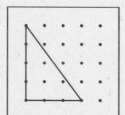

1.

2.

3.

4.

5.

In Exercises 6–8, identify the quadrilateral from its appearance. Then draw another quadrilateral that has the same characteristics.

6.

7.

8.

9. Find the values of x and y for the kite given at the right. Then find the perimeter of the kite.

6.

7.

8.

9.

10.8 Short Quiz

Name_____

Date _____

In Exercises 1 and 2, use the words equilateral, equiangular, and regular to describe the polygon. Then sketch and write measurements for a congruent polygon.

1.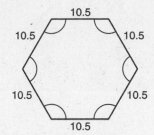

1.

2.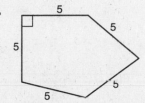

2.

3. Draw an equilateral, equiangular quadrilateral. Identify the figure you drew by name.

3._____

In Exercises 4–7, use the figure.

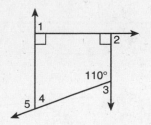

4. What is the sum of the interior angles of the trapezoid?

4._____

5. What is the sum of the exterior angles of the trapezoid?

5._____

6. Does the trapezoid appear to be regular? Explain.

6.

7. Find the measures of angles 1, 2, 3, 4, and 5.

7.

In Exercises 1–4, use the figure. (10.1)

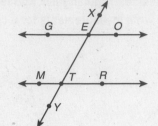

1. Write two other names for line $\overleftrightarrow{MT}$.

1._____

2. Name five line segments that have T as an endpoint.

2._____

3. Write another name for the ray $\overrightarrow{GE}$.

3._____

4. Write three other names for $\angle OEY$.

4._____

In Exercises 5 and 6, use a protractor to construct an angle of the specified measure. Then classify the angle as acute, right, or obtuse. (10.2)

5. 38°

5.

6. 125°

6.

In Exercises 7–12, use the figure. (10.3)

7. Identify the pair of parallel lines.

7._____

8. Name all of the pairs of vertical angles.

8._____

9. List all angles whose measure is 75°.

9._____

10. List all angles whose measure is 112°.

10._____

11. List a pair of corresponding angles that have the same measure.

11._____

12. List a pair of corresponding angles that have different measures.

12._____

In Exercises 13 and 14, classify the triangle by its sides and by its angles. (10.5)

13.

13._____

14.

14._____

15. Sketch a quadrilateral with one pair of congruent sides and one pair of parallel sides. Identify the quadrilateral. (10.6)

15.

In Exercises 16 and 17, use the figure at the right. (10.4, 10.7)

16. Describe the polygon. Then sketch a polygon that is congruent.

16.

17. Identify any symmetry of the figure.

17._____

In Exercises 18–20, use the figure. (10.8, 10.9)

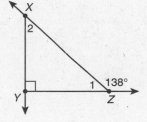

18. What is the measure of angle 1?

18._____

19. What is the measure of angle 2?

19._____

20. Which is the shortest side?

20._____

21. Draw a triangle with lengths of side $a = 6$, side $b = 8$, and side $c = 10$. Then classify the triangle according to its sides and angles. (10.9)

21.

22. Draw a figure that has rotational and line symmetry. Describe the rotational symmetry. (10.4)

22.

23. Draw a regular hexagon. Find the measure of each of its interior angles and exterior angles. (10.8)

23.

24. Order the sides from shortest to longest. (10.9)

24.

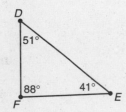

In Exercises 25–27, find the measures of the angles of the triangle. (10.8, 10.9)

25.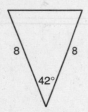

25. _____

26.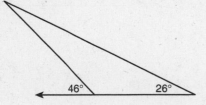

26. _____

27.

27. _____

In Exercises 1–4, use the figure. (10.1)

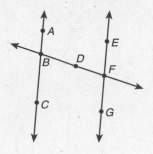

1. What is another name for the line $\overleftrightarrow{AB}$?

   a. $\overleftrightarrow{FG}$ b. $\overleftrightarrow{BD}$

   c. $\overleftrightarrow{BC}$ d. $\overleftrightarrow{EF}$

   1._____

2. Which line segment does NOT have B as an endpoint?

   a. $\overline{BD}$ b. $\overline{BA}$

   c. $\overline{BC}$ d. $\overline{AC}$

   2._____

3. Which is another name for ray $\overrightarrow{FB}$?

   a. $\overrightarrow{FD}$ b. $\overrightarrow{FG}$ c. $\overrightarrow{FE}$ d. $\overrightarrow{FF}$

   3._____

4. Which is NOT another name for $\angle ABF$?

   a. $\angle ABD$ b. $\angle ABC$ c. $\angle FBA$ d. $\angle DBA$

   4._____

In Exercises 5–7, approximate the measure of each angle and classify it. (10.2)

5.   a. 48°, acute   b. 132°, obtuse

   c. 48°, obtuse   d. 132°, acute

   5._____

6. 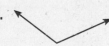   a. 61°, acute   b. 119°, obtuse

   c. 61°, obtuse   d. 119°, acute

   6._____

7.   a. 90°, acute   b. 45°, acute

   c. 90°, right   d. 45°, obtuse

   7._____

8. What symmetry does the figure have, if any? (10.4)

      a. One line of symmetry   b. Two lines of symmetry

   c. No symmetry   d. Rotational and line symmetry

   8._____

9. Which figures are congruent? (10.7)

   | 1 | 2 | 3 | 4 |

   a. 1 and 2   b. 2 and 4   c. 2 and 3   d. 1 and 3

   9._____

In Exercises 10–14, use the figure. (10.3)

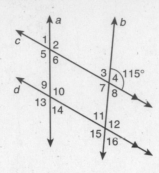

10. Which lines are parallel?   10._____

    a. Lines *a* and *b*   **b.** Lines *c* and *d*

    c. Lines *a* and *c*   **d.** Lines *b* and *d*

11. Which angles are vertical angles?   11._____

    a. ∠s1 and 2   **b.** ∠s1 and 6   **c.** ∠s1 and 3   **d.** ∠s1 and 5

12. Which angle has a measure of 115°?   12._____

    a. ∠2   **b.** ∠3   **c.** ∠7   **d.** ∠8

13. Which are corresponding angles with the same measure?   13._____

    a. ∠s1 and 3   **b.** ∠s1 and 6   **c.** ∠s1 and 9   **d.** ∠s9 and 11

14. Which are corresponding angles with different measures?   14._____

    a. ∠s6 and 14   **b.** ∠s6 and 8   **c.** ∠s9 and 12   **d.** ∠s1 and 4

In Exercises 15–17, classify the figure by its appearance. (10.5, 10.6)

15.   15._____

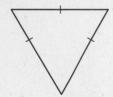

    a. Equilateral equiangular

    b. Acute isosceles

    c. Obtuse isosceles

    d. Obtuse scalene

16.   16._____

    a. Acute scalene

    b. Acute isosceles

    c. Right isosceles

    d. Obtuse scalene

17.   17._____

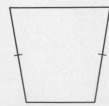

    a. Regular quadrilateral

    b. Isosceles trapezoid

    c. Equilateral rhombus

    d. Kite

In Exercises 18–20, use the figure. (10.8)

18. What is the measure of ∠2?
    a. 120°   b. 60°
    c. 70°   d. 90°

18. _____

19. What is the measure of ∠3?
    a. 120°   b. 60°
    c. 70°   d. 130°

19. _____

20. What is the sum of the interior angles of the figure?
    a. 720°   b. 500°   c. 180°   d. 360°

20. _____

In Exercises 21 and 22, what is the order of the sides, from shortest to longest? (10.9)

21.

a. $\overline{JK}, \overline{KL}, \overline{LJ}$

b. $\overline{KL}, \overline{LJ}, \overline{JK}$

c. $\overline{JL}, \overline{JK}, \overline{KL}$

d. $\overline{JL}, \overline{KL}, \overline{JK}$

21. _____

22.

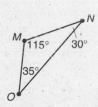

a. $\overline{MN}, \overline{NO}, \overline{OM}$

b. $\overline{MN}, \overline{MO}, \overline{NO}$

c. $\overline{MO}, \overline{MN}, \overline{ON}$

d. $\overline{MO}, \overline{NO}, \overline{MN}$

22. _____

23. What is the measure of each exterior angle of a regular hexagon? (10.8)
    a. 60°   b. 90°   c. 120°   d. 180°

23. _____

In Exercises 24 and 25, what are the measures of the angles of the triangle? (10.8)

24.

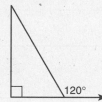

a. 30°, 60°, 90°

b. 45°, 45°, 90°

c. 35°, 55°, 90°

d. 40°, 60°, 90°

24. _____

25.

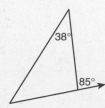

a. 38°, 85°, 57°

b. 38°, 67°, 75°

c. 38°, 95°, 57°

d. 38°, 95°, 47°

25. _____

Chapter 10 Test

Form C
(Page 1 of 3 pages)

Name _____

Date _____

In Exercises 1–7, use the diagram. (10.1–10.3)

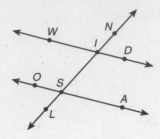

1. Write all the ways $\overleftrightarrow{LN}$ can be named.

   1. _____

2. Name 4 rays that have point I as an endpoint.

   2. _____

3. Name two lines that intersect $\overleftrightarrow{NI}$.

   3. _____

4. Name two lines that do NOT intersect. Identify how these lines are related.

   4.

5. Name two vertical angles that are obtuse.

   5. _____

6. Name two corresponding angles that are acute.

   6. _____

7. Name one straight angle.

   7. _____

In Exercises 8–10, use a protractor. (10.2)

8. Write the measure of an acute angle. Then draw it.

   8. _____

9. Write the measure of an obtuse angle. Then draw it.

   9. _____

10. Write the measure of a right angle. Then draw it.

    10. _____

11. How is a right angle different from an acute angle or an obtuse angle? (10.2)

    11.

Passport to Algebra and Geometry

In Exercises 12 and 13, draw triangles that match the characteristics given. Use markings to show congruent sides. (10.5)

12. Isosceles acute

12.

13. Scalene right

13.

14. Find the values of x and y for the kite given below. (10.6)

14._____

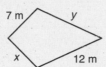

15. Draw a figure with line and rotational symmetry. Draw lines or describe the symmetry. (10.4)

15.

16. Divide the figure into as few congruent triangles as possible.

16.

17. Order the angles of the triangle from smallest to largest. (10.9)

17._____

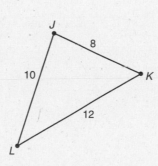

In Exercises 18–22, use the figure. (10.3)

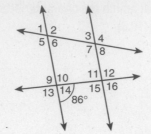

18. Name two pairs of vertical angles.

18._____

19. Which angle is the non-congruent corresponding angle to angle 9?

19._____

20. Which angle is the congruent corresponding angle to angle 16? What is its measure?

20._____

21. Which is greater: $m\angle 13$ or $m\angle 14$? Explain.

21.

22. If the $m\angle 7$ is 107°, which other angles have the same measure?

22._____

In Exercises 23–27, use the figure. (10.8)

23. What is the measure of angle 1?

23._____

24. What is the measure of angle 4?

24._____

25. What is the measure of angle 3?

25._____

26. What is the measure of angle 2?

26._____

27. What is the sum of the measures of the interior angles?

27._____

In Exercises 28 and 29, find the measures of angles 1 and 2 of the triangles. Then order the sides from shortest to longest. (10.8, 10.9)

28.

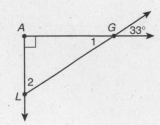

28.

29.

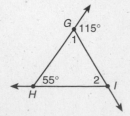

29.

Passport to Algebra and Geometry

Name_____

Date _____

In Exercises 1 and 2, use the figures.

 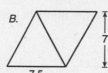

1. Find the area of figure A.

2. Describe two ways to find the area of figure B. Then write the area.

3. Make a dot grid sketch the represent the "equation" (1 rectangle) + (2 triangles) = (1 trapezoid). Then find the area and perimeter of the figure.

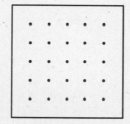

4. Estimate the area of the figure if 1 square unit = 50 square meters.

1._____

2._____

3._____

4._____

In Exercises 5–10, use the fact that △TOP ≅ △LID to complete the statement.

5. ∠T ≅ _____

6. $\overline{TO}$ ≅ _____

7. $\overline{LD}$ ≅ _____

8. ∠I ≅ _____

9. ∠P ≅ _____

10. $\overline{OP}$ ≅ _____

11. Draw lines to divide the figure into two congruent parts. Give more than one answer, if possible.

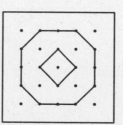

5._____

6._____

7._____

8._____

9._____

10._____

11.

In Exercises 1 and 2, draw the reflection of the figure in line a.

1.

1.

2.

2.

3. Sketch the trapezoid in the coordinate plane after it is reflected about the y-axis.

3.

4. Which digits, 0–9, remain the same when reflected about a horizontal line? a vertical line?

4._____

5. The figure in Quadrant 1 is rotated clockwise about the origin to produce the figure in Quadrant 3. Find the angle of rotation.

5._____

6. Estimate the angle and the direction of rotation for the figures at the right.

6._____

In Exercises 7–10, △RST is rotated 90° clockwise about the origin to produce △R′S′T′.

7. Draw △R′S′T′.

7._____

8. What is the length of $\overline{R'T'}$?

8._____

9. Which angle of △R′S′T′ is a right angle?

9._____

10. Find the area of △R′S′T′.

10._____

Name _____

Date _____

1. Draw and name a polygon for which the area can be expressed as (base)(height). (11.1)

1. _____

In Exercises 2 and 3, use the figure. (11.1)

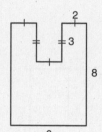

2. Find the area of the figure.

2. _____

3. Find the perimeter of the figure.

3. _____

In Exercises 4–7, use the congruent triangles to complete the statement. (11.2)

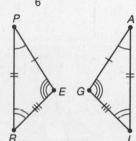

4. $\angle A \cong$ _____

4. _____

5. $\overline{PR} \cong$ _____

5. _____

6. $\angle G \cong$ _____

6. _____

7. $\overline{LG} \cong$ _____

7. _____

8. Identify the line of reflection. (11.3)

8. _____

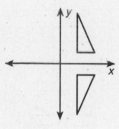

9. Identify the angle and direction of the rotation. (11.4)

9. _____

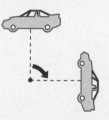

10. Write a verbal description of the translation. (11.5)

10.

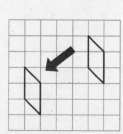

Name_____

Date _____

1. Draw and name a polygon for which
   the area can be expressed as
   $\frac{1}{2}$ (base)(height). (11.1)

1. _____

In Exercises 2 and 3, use the figure. (11.1)

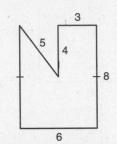

2. Find the area of the figure.

2. _____

3. Find the perimeter of the figure.

3. _____

**In Exercises 4–7, use the congruent
triangles to complete the statement.**

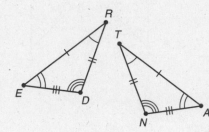

4. $\angle A \cong$ ____

4. _____

5. $\overline{ED} \cong$ ____

5. _____

6. $\angle R \cong$ ____

6. _____

7. $\overline{TN} \cong$ ____

7. _____

8. Identify the line of reflection. (11.3)

8. _____

9. Identify the angle and direction of the
   rotation. (11.4)

9. _____

10. Write a verbal description of the
    translation. (11.5)

10.

Passport to Algebra and Geometry

In Exercises 1 and 2, describe the translation verbally. Then write the ordered pair that describes the translation.

1.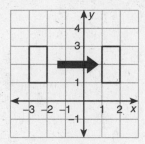

1. _____

2.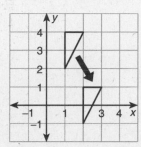

2. _____

3. Draw a triangle whose vertices are (3, 2), (6, 2), and (6, 5). Then translate the triangle 2 units to the left and 1 unit down.

3.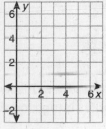

4. Draw a figure that is similar.

4. _____

In Exercises 5–8, trapezoids TRAP and ZOID are similar, as shown.

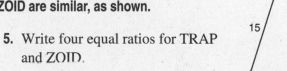

5. Write four equal ratios for TRAP and ZOID.

5. _____

6. Find the scale factor of TRAP to ZOID.

6. _____

7. Find the length of *AP*.

7. _____

8. ∠*D* ≅ _____

8. _____

Name_____

Date _____

1. The scale drawing of a rectangular park has a scale factor of 1 cm
   to 74 m. The drawing is 11 cm by 18 cm. What are the actual dimensions of the park?

1.

2. A rectangular room is 20 feet long by 28 feet wide. Determine your
   own scale for drawing a model of the room. Then draw and label the
   model using your scale.

2.

**In Exercises 3–5, use △CLS to find the
trigonometric ratio.**

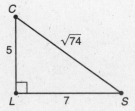

3. sin C

3._____

4. cos C

4._____

5. tan C

5._____

**In Exercises 6 and 7, solve the triangle for its unlabeled angle and side.
Then write six trigonometric ratios that can be formed with the triangle.**

6.

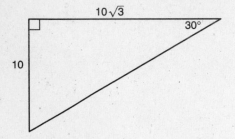

6.

7.

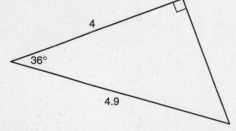

7.

In Exercises 1 and 2, find the perimeter and area of the figure. (11.1)

1.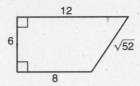

1._____

2.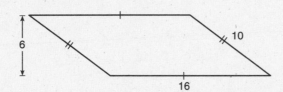

2._____

In Exercises 3–8, use the fact that $\triangle BOX \cong \triangle TIN$ to complete the statement. (11.2)

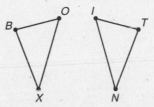

3. $\overline{BO} \cong$ _____

3._____

4. $\overline{TN} \cong$ _____

4._____

5. $\overline{XO} \cong$ _____

5._____

6. $\angle B \cong$ _____

6._____

7. $\angle O \cong$ _____

7._____

8. $\angle N \cong$ _____

8._____

In Exercises 9–12, find the vertices of the figure after performing the indicated transformation. (11.3–11.5)

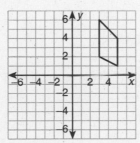

9. Reflect the figure about the *x*-axis.

9._____

10. Rotate the figure clockwise 90° about the origin.

10._____

11. Translate the figure 8 units to the left and 6 units down.

11._____

12. Reflect the figure about the *y*-axis, then translate it 1 unit to the right.

12._____

In Exercises 13–15, describe the transformation that will map one smiling face to the other. (11.3–11.5)

13.

13. _____

14.

14. _____

15.

15. _____

In Exercises 16–19, use the fact that
△*NYC* ~ △*LAX*. **(11.6)**

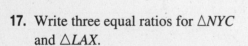

16. Find *NC* and *LA*.

16. _____

17. Write three equal ratios for △*NYC*
and △*LAX*.

17. _____

18. Find the scale factor of △*NYC* to △*LAX*.

18. _____

19. Find the ratio of the perimeter of △*NYC* to the perimeter of △*LAX*.
Is this ratio equal to the scale factor?

19. _____

20. Draw a trapezoid that is similar to the one
shown. Label the lengths of each side and
give the scale factor. (11.6)

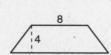

20. _____

In Exercises 21–24, use the scale drawing of the picture frame. (11.7)

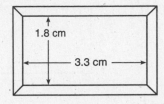

1.8 cm

3.3 cm

21. The actual height of the inside of the picture frame is 13.5 centimeters. Use the scale drawing to find the actual width.

21._____

22. Find the scale factor of the actual frame to the scale drawing.

22._____

23. Find the area of the frame and the scale drawing.

23._____

24. Find the ratio of the area of the actual frame to the area of the scale drawing. How does this relate to the scale factor?

24._____

In Exercises 25–30, use △SAM to find the trigonometric ratio. Round your result to two decimal places. (11.8)

S ——10—— A

26

24

M

25. sin S

25._____

26. cos S

26._____

27. tan S

27._____

28. sin M

28._____

29. cos M

29._____

30. tan M

30._____

31. Find the value of x. (11.9)

31._____

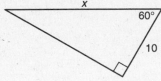

x

60°

10

In Exercises 1 and 2, use the polygon. (11.1)

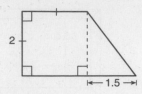

1. What is the perimeter of the polygon?

   a. 10 units   **b.** 7 units   **c.** 11 units   **d.** 10.5 units

   1._____

2. What is the area of the polygon?

   a. 7 square units   **b.** 11 square units

   c. 10.5 square units   **d.** 5.5 square units

   2._____

In Exercises 3–8, use the fact that
△**BIG** ≅ △**TAL. (11.2)**

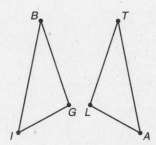

3. To what is $\overline{BI}$ congruent?

   a. $\overline{IG}$ **b.** $\overline{GB}$ **c.** $\overline{TA}$ **d.** $\overline{AL}$

   3._____

4. To what is $\overline{IG}$ congruent?

   a. $\overline{BG}$ **b.** $\overline{TA}$ **c.** $\overline{LA}$ **d.** $\overline{LT}$

   4._____

5. To what is $\overline{GB}$ congruent?

   a. $\overline{BI}$ **b.** $\overline{AT}$ **c.** $\overline{AL}$ **d.** $\overline{LT}$

   5._____

6. To what is ∠B congruent?

   a. ∠I   **b.** ∠G   **c.** ∠T   **d.** ∠A

   6._____

7. To what is ∠A congruent?

   a. ∠L   **b.** ∠T   **c.** ∠B   **d.** ∠I

   7._____

8. To what is ∠G congruent?

   a. ∠L   **b.** ∠B   **c.** ∠T   **d.** ∠A

   8._____

9. Which statement is always true? (11.6)

   9._____

   a. Two similar figures are always congruent.

   b. Two congruent figures are always similar.

   c. A rectangle and a square are never congruent.

   d. A rhombus and a square are never similar.

In Exercises 10–13, what is the coordinate of point X after performing the indicated transformation? (11.3–11.5)

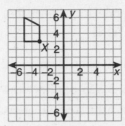

10. Reflect the figure about the *x*-axis.

   a. $(-3, -3)$ **b.** $(3, 3)$ **c.** $(-3, 3)$ **d.** $(3, -3)$

10._____

11. Rotate the figure 180° clockwise about the origin.

   a. $(3, -3)$ **b.** $(-3, 0)$ **c.** $(3, 0)$ **d.** $(-3, 3)$

11._____

12. Translate the figure 4 units to the right and 1 unit up.

   a. $(4, 1)$ **b.** $(1, 4)$ **c.** $(-2, 7)$ **d.** $(2, -7)$

12._____

13. Reflect the figure in the *y*-axis, then translate it 2 units down.

   a. $(3, 1)$ **b.** $(1, 3)$ **c.** $(-3, -1)$ **d.** $(3, -1)$

13._____

In Exercises 14–17, is the shaded house congruent to the white house? If so, what transformation maps the shaded house to the white house? (11.3–11.5)

14.

   a. reflection over a line

   b. rotated 135° clockwise

   c. moved 3 units to the right and 2 units down

   d. not congruent

14._____

15.

   a. reflection over a line

   b. rotated 180°

   c. moved 5 units to the left and 2 units up

   d. not congruent

15._____

16.

   a. reflection over a line

   b. rotated 90°

   c. moved 2 units to the left and 2 units up

   d. not congruent

16._____

17.

   a. reflection over a line

   b. rotated 150°

   c. moved 3 units to the right and 3 units down

   d. not congruent

17._____

In Exercises 18 and 19, you are building a model for a bench that is 153 centimeters wide and 72 centimeters high. (11.7)

18. If the model is 18 centimeters high, how wide should it be?

    a. 38 centimeters   b. 2.125 centimeters

    c. 38.25 centimeters   d. 21.25 centimeters

18._____

19. What would the scale factor be of the model to the real bench?

    a. 4   b. 0.25   c. 18   d. 2.125

19._____

In Exercises 20–25, use △REF to find the trigonometric ratio. Round your result to two decimal places. (11.8)

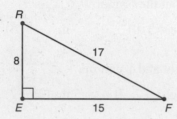

20. What is sin R?

    a. 0.88   b. 0.12   c. 0.15   d. 1.13

20._____

21. What is sin F?

    a. 2.125   b. 0.47   c. 0.88   d. 0.53

21._____

22. What is cos R?

    a. 0.47   b. 0.53   c. 0.12   d. 0.88

22._____

23. What is cos F?

    a. 0.47   b. 0.12   c. 0.53   d. 0.88

23._____

24. What is tan R?

    a. 1.88   b. 2.13   c. 0.53   d. 0.47

24._____

25. What is tan F?

    a. 1.88   b. 2.13   c. 0.53   d. 0.47

25._____

26. What is the value of x? Round your answer to the nearest whole number. (11.9)

    a. 20   b. 17

    c. 23   d. 15

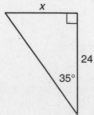

26._____

In Exercises 1 and 2, find the perimeter and area of the figure. (11.1)

1.

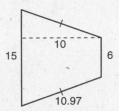

1. _____

2.

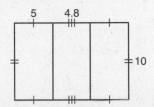

2. _____

In Exercises 3 and 4, use the fact that
△CAR ≅ △VAN. (11.2)

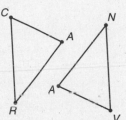

3. Write three statements showing the congruence of the sides.

3. _____

4. Write three statements showing the congruence of the angles

4. _____

In Exercises 5–10, use the figure.
(11.3–11.5)

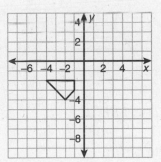

5. Write the coordinates for the vertices of the figure.

5. _____

6. Reflect the figure about the x-axis. Write the coordinates of the image.

6. _____

7. Reflect the figure about the y-axis. Write the coordinates of the image.

7. _____

8. Describe the relationship between the coordinates of the original figure and the coordinates of the images from the above reflections.

8.

9. Write the coordinates for the vertices if the figure is transformed 7 units to the right and 6 units down.

9. _____

10. Write the coordinates for the vertices if the figure is rotated 180° in a clockwise direction.

10. _____

In Exercises 11–13, state whether the white kite is congruent to the shaded kite. If it is, describe the transformation that will map the white kite to the shaded kite. (11.3–11.5)

11.

11._____

12.

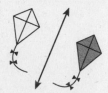

12._____

13.

13._____

In Exercises 14–20, use the fact that
ABCD ~ EFGH. (10.6)

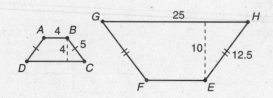

14. Write four equal ratios for *ABCD* and *EFGH*.

14.

15. Find the scale factor for *ABCD* to *EFGH*.

15._____

16. Find *EF*.

16._____

17. Find *CD*.

17._____

18. Find the ratio of the perimeter of *ABCD* to the perimeter of *EFGH*.

18._____

19. Find the ratio of the area of *ABCD* to the area of *EFGH*.

19._____

20. Explain the relationship between the ratios of the sides, perimeters, and areas of *ABCD* to *EFGH*.

20._____

In Exercises 21–28, use △*TOP*. Round results to 2 decimal places, if necessary. (11.8)

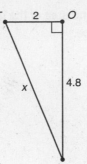

21. Find the measure of the third side.

21._____

22. Find sin *T*.

22._____

23. Find cos *T*.

23._____

24. Find tan *T*.

24._____

25. Find sin *P*.

25._____

26. Find cos *P*.

26._____

27. Find tan *P*.

27._____

28. Find two trigonometric ratios above that are the same. Explain why they are the same.

28._____

In Exercises 29 and 30, find the value of *x* and *y*. Round your answers to the nearest tenth. (11.9)

29.

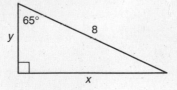

29._____

30.

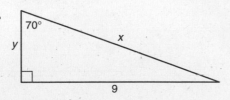

30._____

In Exercises 31 and 32, use the picture. (11.9)

31. How high in the tree is the kitten?

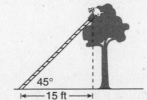

31._____

32. Find the length of the ladder.

32._____

Name_____

Date _____

In Exercises 1 and 2, find the circumference and area of the figure. Use 3.14 for π. Round your answer to one decimal place.

1.
   r = 5 ft

1._____

2.
   d = 4.2 in.

2._____

In Exercises 3 and 4, find the radius and diameter of the figure. Use 3.14 for π. Round your answer to one decimal place.

3.
   C = 16.96 cm

3._____

4.
   A = 36.32 m

4._____

5. Find the area of the shaded portion of the figure. Use 3.14 for π. Round your answer to one decimal place.

d = 12 m

5._____

In Exercises 6–9, name an object for each solid.

6. Sphere

6._____

7. Prism

7._____

8. Cylinder

8._____

9. Cone

9._____

10. Which solid cannot be formed by folding a net? Explain.

10.

Name_____

Date _____

In Exercises 1 and 2, find the surface area of the solid. Use 3.14 for π.

1.

1._____

2.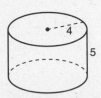

2._____

3. A cube has a surface area of 150 centimeters. Draw the cube showing the measurement of each side.

3.

4. Find the volume of the figure.

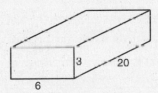

4._____

5. The volume is 120 cm³. Find the value of x.

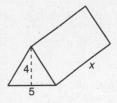

5._____

In Exercises 6–8, a shoe box is 11 inches long, 5 inches wide, and 4 inches high.

6. Sketch the shoe box.

6.

7. How many one-inch cubes could be stored in the shoe box?

7._____

8. If you wanted to wrap the shoe box, how much wrapping paper would you need?

8._____

In Exercises 1 and 2, use 3.14 for π. (12.1)

1. Find the area and the circumference of the circle.

$d = 18$ ft

1._____

2. Find the area of the shaded region.

$r = 8$ cm

2._____

In Exercises 3 and 4, find the surface area of the solid. Use 3.14 for π. (12.3)

3.

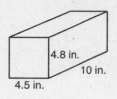

4.8 in.

10 in.

4.5 in.

3._____

4.

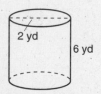

2 yd

6 yd

4._____

5. Draw a cube. Label a face, an edge, a vertex, and a base. (12.2)

5.

In Exercises 6 and 7, Olu and Rachel are building a mesh framework for a compost pile. The framework encloses an area that is 12 feet long and 8 feet wide. The volume of the compost pile when filled to capacity is 576 ft³. (12.4)

6. What is the height of the framework?

6._____

7. Sketch the completed framework and label each dimension.

7.

In Exercises 1 and 2, use 3.14 for π. (12.1)

1. Find the area and the circumference
   of the circle.

d = 22 yd

1._____

2. Find the area of the shaded region.

d = 20 ft

2._____

In Exercises 3 and 4, find the surface area of the solid. Use 3.14 for π. (12.3)

3.

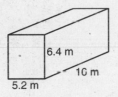
6.4 m
10 m
5.2 m

3._____

4.

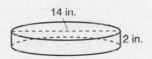
14 in.
2 in.

4._____

5. Draw a rectangular prism. Label a face, an edge, a vertex,
   and a base. (12.2)

5.

In Exercises 6 and 7, Mrs. Fleming bought a freezer. It is 4 feet long, 2 feet wide, and has a volume of 36 cubic feet. (12.4)

6. Find the freezer's height.

6._____

7. Sketch the completed freezer and label each dimension.

7.

Name_____

Date_____

1. Find the volume of the cylinder. Use 3.14 for π.

1._____

In Exercises 2 and 3, find the height or the radius of the base. Use 3.14 for π. Round your answer to two decimal places.

2.

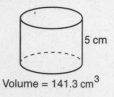

Volume = 141.3 cm^3

2._____

3.

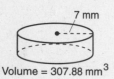

Volume = 307.88 mm^3

3._____

In Exercises 4 and 5, find the volume of the figure. Use 3.14 for π.

4.

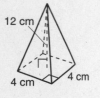

4._____

5.

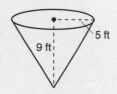

5._____

6. Draw a figure with a base radius of 6 cm and a height of 10 cm. Find the volume. Use 3.14 for π.

6._____

12.8 Short Quiz

Name _____

Date _____

In Exercises 1 and 2, find the volume. Use 3.14 for π. Round your results to 2 decimal places.

1.

r = 7 in.

1. _____

2.

r = 8.1 cm

2. _____

3. Find the radius of a sphere with a volume of 33.51 cubic feet.

3. _____

4. Find the radius of a sphere with a volume of 288 π cubic inches.

4. _____

In Exercises 5 and 6, sketch a solid that is similar. Write the dimensions of the solid.

5.

2.5 in.

10 in.

5.

6.

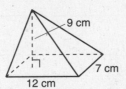

9 cm

7 cm

12 cm

6.

7. Use a proportion to solve for x and y in the similar solids.

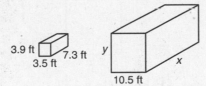

3.9 ft

3.5 ft

7.3 ft

y

x

10.5 ft

7. _____

8. Name a solid that is always similar to another like solid.

8. _____

9. Name a solid that may or may not be similar to another like solid. Explain.

9.

In Exercises 1 and 2, find the circumference and area of the circle. Use 3.14 for π. (12.1)

1.

1._____

2.

2._____

In Exercises 3–7, use the net. Use 3.14 for π. Round your results to 2 decimal places. (12.1, 12.4, 12.5)

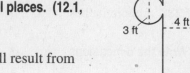

3. Describe the solid that will result from folding the net.

3._____

4. Find the circumference of one of the bases.

4._____

5. Find the area of one of the bases.

5._____

6. Find the surface area of the solid.

6._____

7. Find the volume of the solid.

7._____

In Exercises 8–11, use the solid figure. (12.2, 12.4)

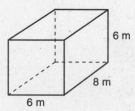

8. Identify the solid.

8._____

9. How many faces, vertices, and edges does the solid have?

9._____

10. Find the surface area of the solid.

10._____

11. Describe how you calculated the surface area.

11._____

In Exercises 12–14, find the surface area. (12.4)

12.

12._____

13.

13._____

14.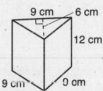

14._____

In Exercises 15–18, find the volume. Use 3.14 for π. (12.5–12.7)

15.

15._____

16.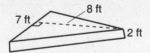

16._____

17.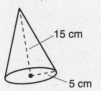

17._____

18.

18._____

19. Decide whether the solids are similar. If so, find the scale factor. (12.8)

19. _____

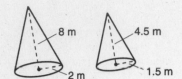

In Exercises 20–22, find the value of *x*. (12.1, 12.3, 12.4)

20.

Area = 314 m²

20. _____

21.

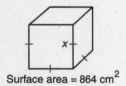

Surface area = 864 cm²

21. _____

22.

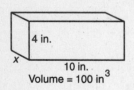

4 in.

x 10 in.
Volume = 100 in³

22. _____

In Exercises 23–27, use the similar solids.
Use 3.14 for π. (12.6, 12.8)

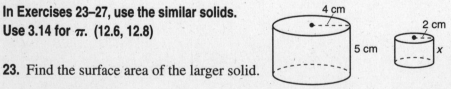

4 cm

2 cm

5 cm   *x*

23. Find the surface area of the larger solid.

23. _____

24. Find the volume of the larger solid.

24. _____

25. Find the scale factor of the small solid to the large solid.

25. _____

26. Find the surface area of the smaller solid.

26. _____

27. Find the volume of the smaller solid.

27. _____

In Exercises 1 and 2, use the circle at the right and 3.14 for π. (12.1)

9 in.

1. What is the circumference of the circle, rounded to two decimal places?

   a. 56 inches   **b.** 56.52 inches   **c.** 5.65 inches   **d.** 565.65 inches

1._____

2. What is the area of the circle, rounded to two decimal places?

   a. 254.34 in.^{2} **b.** 25.434 in.^{2} **c.** 254.00 in.^{2} **d.** 2543.40 in.^{2}

2._____

In Exercises 3 and 4, use the circle at the right and 3.14 for π. (12.1)

$C = 69.08$ ft

3. What is the radius of the circle?

   a. 23 feet   **b.** 22 feet   **c.** 11 feet   **d.** 11.5 feet

3._____

4. What is the area of the circle, rounded to the nearest whole number?

   a. 38 feet   **b.** 380 feet   **c.** 4761 feet   **d.** 217 feet

4._____

In Exercises 5–9, use the solid figure. (12.2, 12.3)

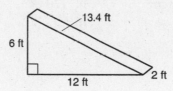

13.4 ft

6 ft

12 ft

2 ft

5. What is the name of this solid?

   a. pyramid   **b.** rectangular prism

   c. triangular prism   **d.** triangular pyramid

5._____

6. How many faces does the figure have?

   a. 3   **b.** 6   **c.** 4   **d.** 5

6._____

7. How many edges does the figure have?

   a. 7   **b.** 5   **c.** 6   **d.** 9

7._____

8. How many vertices does the figure have?

   a. 5   **b.** 6   **c.** 7   **d.** 8

8._____

9. What is the surface area of this figure?

   a. 134.8 ft^2 **b.** 74.8 ft^2 **c.** 1929.6 ft^2 **d.** 964.8 ft^2

9._____

In Exercises 10–13, use the cylinder. Use 3.14 for π. (12.1, 12.3, 12.4)

15 cm

3 cm

10. What is the circumference of one of the bases?

10._____

a. 18.84 centimeters

b. 45 centimeters

c. 9.42 centimeters

d. 28.27 centimeters

11. What is the area of one of the bases?

11._____

a. 18.84 cm²   b. 45 cm²   c. 9.42 cm²   d. 28.26 cm²

12. What is the surface area of the cylinder?

12._____

a. 45 cm²   b. 282.6 cm²   c. 424.05 cm²   d. 339.12 cm²

13. What is the volume of the cylinder?

13._____

a. 45 cm³   b. 282.6 cm³   c. 423.9 cm³   d. 339.14 cm³

In Exercises 14 and 15, what is the surface area of the solid? Use 3.14 for π. (12.3, 12.4)

14.

5 m

6 m

4 m

a. 148 m²   b. 120 m²

c. 15 m²   d. 74 m²

14._____

15.

8 mm

1.5 mm

a. 37.7 mm²   b. 125.66 mm²

c. 138.16 mm²   d. 301.59 mm²

15._____

In Exercises 16 and 17, what solid will result from folding the net? (12.2)

16.

a. rectangular prism

b. rectangular pyramid

c. triangular pyramid

d. triangular prism

16._____

17.

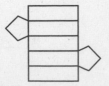

a. hexagonal prism

b. pentagonal pyramid

c. cone

d. pentagonal prism

17._____

In Exercises 18–21, find the volume. Use 3.14 for π. (12.5–12.7)

18.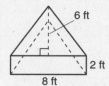

a. 64 ft³   b. 48 ft³
c. 96 ft³   d. 32 ft³

18._____

19.

a. 14,130 ft³   b. 942 in.³
c. 3769 in.³   d. 7539 in.³

19._____

20.

a. 60 cm³   b. 720 cm³
c. 314 cm³   d. 628 cm³

20._____

21.

a. 96 in.³   b. 64 in.³
c. 192 in.³   d. 32 in.³

21._____

In Exercises 22–25, use the similar figures at the right. (12.5, 12.8)

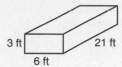
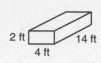

22. Find the scale factor of the larger prism to the smaller prism.

a. 2   b. 1.5   c. 2.5   d. 3

22._____

23. Find the volume of the larger prism.

a. 189 ft³   b. 225 ft³   c. 378 ft³   d. 414 ft³

23._____

24. Find the volume of the smaller prism.

a. 224 ft³   b. 56 ft³   c. 240 ft³   d. 112 ft³

24._____

25. Find the scale factor of the volume of the larger prism to the smaller prism.

a. 3.375   b. 2.25   c. 1.5   d. 1.725

25._____

In Exercises 1 and 2, use 3.14 for π. Round your results to 2 decimal places. (12.1)

1. Find the circumference and area of the circle.

14.8 m

1. _____

2. Find the area of the shaded region.

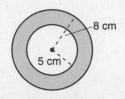

8 cm

5 cm

2. _____

In Exercises 3–5, a circular fountain in the downtown shopping mall has an area of 379.94 ft². Use 3.14 for π. (12.4)

3. Find the circumference of the fountain.

3. _____

4. The water shoots up from a pipe in the center of the fountain. How far is the center of the pipe from any point on the fountain's outside rim?

4. _____

5. Sketch the fountain. Show the position of the water pipe and label its distance from a point on the outside rim.

5. _____

In Exercises 6–12, use the solid figure.
(12.2, 12.3, 12.5)

6. Identify the solid figure.

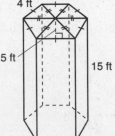

4 ft

5 ft

15 ft

6. _____

7. How many faces does the solid have?

7. _____

8. How many edges does the solid have?

8. _____

9. How many vertices does the solid have?

9. _____

10. What is the area of one of the bases?

10. _____

11. What is the surface area of the figure?

11. _____

12. What is the volume of the figure?

12. _____

13. A polyhedron has one square face and 4 triangular faces. Identify
the polyhedron and sketch its net. (12.2)

13. _____

**In Exercises 14–16, find the surface area of the solid. Then describe the
procedure you used to find the answer. Use 3.14 for π. (12.3, 12.4)**

14.

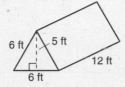

6 ft   5 ft   12 ft   6 ft

14. _____

15.

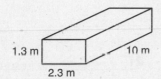

1.3 m   10 m   2.3 m

15. _____

16.

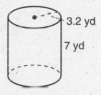

3.2 yd   7 yd

16. _____

**In Exercises 17 and 18, sketch a figure that is similar to the figure shown.
Write dimensions for your figure, using the scale factor given. (12.8)**

17.

15.6 cm   5.2 cm

Scale factor: 4

17. _____

18.

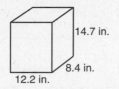

14.7 in.   8.4 in.   12.2 in.

Scale factor: $\frac{2}{3}$

18. _____

19. Write dimensions for the second figure shown using a scale factor of your choice. Identify the scale factor you used to calculate the dimensions. (12.8)

19._____

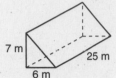 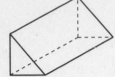

7 m 25 m 6 m

In Exercises 20–22, find the volume. Use 3.14 for π. Round your results to 2 decimal places. (12.5, 12.6)

20.

10 m
2.5 m

20._____

21.

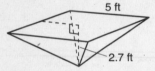

5 ft
2.7 ft

21._____

22.

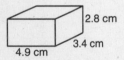

2.8 cm
3.4 cm
4.9 cm

22._____

23. The Great Pyramid of Egypt has a square base. Each side of the base is 230 meters in length. The height is 147 meters. Sketch the Great Pyramid. Then find its volume. (12.6)

23._____

In Exercises 24–27, use the similar spheres. Use 3.14 for π. (12.7, 12.8)

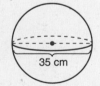

5 cm
35 cm

24. Find the scale factor of Figure A to Figure B.

24._____

25. Find the volume of the larger sphere.

25._____

26. Find the volume of the smaller sphere.

26._____

27. Find the scale factor of the volume of Figure A to Figure B. Explain why your answer is reasonable.

27._____

Passport to Algebra and Geometry

**In Exercises 1–4, evaluate the expression. Then simplify, if possible.
(7.1, 7.2, 7.4, 7.5)**

1. $\frac{3}{8} + \frac{5}{6}$

    1. _____

2. $\frac{6}{7} - \frac{3}{14}$

    2. _____

3. $\frac{3}{5} \cdot \frac{9}{10}$

    3. _____

4. $\frac{2}{3} \div \frac{4}{5}$

    4. _____

In Exercises 5–8, solve the equation.   (7.1, 7.2, 7.4, 7.5)

5. $x - \frac{1}{2} = \frac{5}{6}$

    5. _____

6. $b + \frac{2}{5} = 1\frac{1}{3}$

    6. _____

7. $4e = \frac{3}{7}$

    7. _____

8. $\frac{3}{5}h = \frac{3}{10}$

    8. _____

**In Exercises 9–12, use a calculator to evaluate the expression. Round your
result to two decimal places.   (7.1–7.5)**

9. $\frac{34}{45} + \frac{13}{14}$

    9. _____

10. $\frac{25}{31} \cdot \frac{12}{21}$

    10. _____

11. $\frac{16}{17} - \frac{6}{7}$

    11. _____

12. $\frac{21}{25} \div \frac{5}{9}$

    12. _____

In Exercises 13–15, write the fraction that represents the portion of the figure's area that is shaded. Then write the fraction as a percent. (7.6)

13.

13. _____

14.

14. _____

15.

15. _____

In Exercises 16 and 17, use the circle graph. (7.7–7.9)

16. Write each percent as a fraction.

17. Four hundred people participated in this survey. How many people answered in each category?

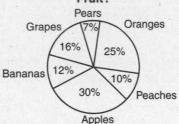

What is Your Favorite Fruit?

16.

17. _____

In Exercises 18 and 19, write the verbal phrase as a rate or a ratio. State whether it is a rate or a ratio. (8.1)

18. 25 students to sit at 5 tables

18. _____

19. 11 out of 20 illustrations in color

19. _____

In Exercises 20–23, solve the proportion. (8.2)

20. $\dfrac{5}{7} = \dfrac{x}{49}$

20._____

21. $\dfrac{12}{5} = \dfrac{108}{y}$

21._____

22. $\dfrac{4}{5} = \dfrac{3}{z}$

22._____

23. $\dfrac{2}{3} = \dfrac{x}{2}$

23._____

In Exercises 24 and 25, decide whether the change is an increase or a decrease. Then find the percent. Round results to 1 decimal place. (8.6)

24. Beginning balance: $345.00
End balance: $500.00

24._____

25. 50 m freestyle on 7/1: 45.2 sec
50 m freestyle on 7/15: 39.8 sec

25._____

In Exercises 26 and 27, you have markers in a case that are the following colors: 3 fine line red, 2 fine line blue, 4 fine line green, and 1 broad tip red. (8.7)

26. If you randomly pick one writing instrument, what is the probability that it is a broad-tip marker?

26._____

27. If you randomly pick one writing instrument, what is the probability that it is green?

27._____

In Exercises 28–30, solve the equation and decide whether the solution is rational or irrational. (9.1, 9.2)

28. $t^2 = 169$

28._____

29. $3r^2 = 15$

29._____

30. $42 + e^2 = 91$

30._____

In Exercises 31–33, *a* and *b* are the lengths of the legs of a right triangle, and *c* is the length of the hypotenuse. Find the missing length. Round your results to 2 decimal places. (9.3)

31. $a = 9, b = 15$

31._____

32. $a = 10, c = 20$

32._____

33. $b = 36, c = 39$

33._____

In Exercises 34–38, solve the inequality. Then graph the solution on a number line. (9.5–9.7)

34. $-9m + 14 < -22$

34.

35. $5s - 1 \geq 29$

35.

36. $2(k + 4) \leq 10$

36.

37. $\dfrac{c}{4} + 7 < -6$

37.

38. $-4 < -4(f - 7)$

38.

In Exercises 39–41, decide whether the triangle can have the given side lengths. Explain. (9.3, 9.8)

39.

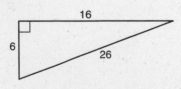

39.

40.

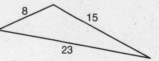

40.

41.

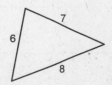

41.

Passport to Algebra and Geometry

In Exercises 42–49, use the figure.
(10.1–10.3)

42. Write two other names for $\overleftrightarrow{SU}$.

43. List 5 line segments that have R as an endpoint.

44. Write another name for $\overrightarrow{LE}$.

45. Write three other names for $\angle LRU$.

46. List two pairs of vertical angles.

47. List 4 angles whose measure is 92°.

48. List two pairs of congruent corresponding angles.

49. List two pairs of noncongruent corresponding angles.

42._____

43._____

44._____

45._____

46._____

47._____

48._____

49._____

In Exercises 50–53, construct an angle of the given length. Then identify it as acute, obtuse, right or straight. (10.2)

50. 90°

51. 43°

52. 180°

53. 157°

50._____

51._____

52._____

53._____

54. Draw lines to identify line symmetry in the figure. Then identify and explain any other kinds of symmetry. (10.4)

54.

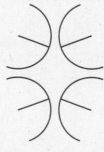

In Exercises 55–57, identify the polygon. (Be specific.) Then find the value of *x*. (10.5, 10.6, 10.8)

55.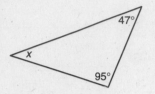

55. _____

56.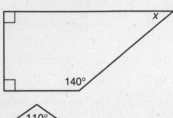

56. _____

57.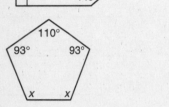

57. _____

In Exercises 58 and 59, order the angles from smallest to largest. (10.9)

58.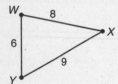

58. _____

59.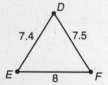

59. _____

In Exercises 60–61, find the perimeter and area of the polygon. Round results to 2 decimal places. (11.1)

60.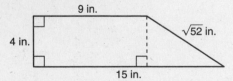

60. _____

61.

61. _____

Passport to Algebra and Geometry

In Exercises 62 and 63, use the figure where △*ABC* is reflected in line *m*. (11.3)

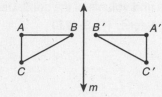

62. Is △*ABC* congruent to △*A′B′C′*?

62. _____

63. Compare the distance between the line *m* and points *A* and *A′*.

63. _____

64. Find the circumference and area of the circle. Use 3.14 for π. (12.1)

64. _____

7 ft

65. Find the radius and the diameter of the circle. Use 3.14 for π. Round your results to the nearest whole number. (12.1)

65. _____

Area = 50 in.^{2}

In Exercises 66 and 67, use Figure A. (11.6)

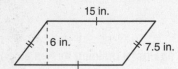

15 in.

6 in.

7.5 in.

66. Sketch a polygon similar to Figure A. Make up your own scale factor and give the side lengths for your figure. Find the perimeters for both figures.

66.

67. Find the area for both figures. Explain the relationship between the two numbers.

67.

In Exercises 68–69, find the surface area and volume of the solid. Use 3.14 for π. Round your result to two decimal places. (12.3–12.5)

68.

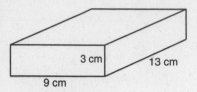

3 cm   13 cm
9 cm

68._____

69.

2 cm
5 cm

69._____

In Exercises 70 and 71, find the volume. Use 3.14 for π. (12.6, 12.7)

70.

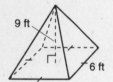

9 ft
6 ft

70._____

71.

1 ft

71._____

In Exercises 72 and 73, a ladder leaning against a building makes an angle of 50° with the ground. The top of the ladder reaches a window 7 meters above the ground. (11.8, 11.9)

72. Draw a sketch showing the situation. Label all the sides and angles you know.

72.

73. Find the length of the ladder. Round your answer to the nearest tenth.

73._____

Passport to Algebra and Geometry

13.2 Short Quiz

Name_____

Date _____

In Exercises 1–4, decide whether the ordered pair is a solution of
$y = 3x + 2$. If it is not a solution, provide the correct y-coordinate for
the given x-coordinate.

1. $(-1, -1)$

2. $(0, 4)$

3. $(1, 6)$

4. $(2, 8)$

5. Find several solutions of the linear equation $2x + 1$. Use a table of
   values to organize your results.

x						
y						

In Exercises 6 and 7, use the fact that one kilogram equals 2.2 pounds.

6. Write an equation to describe the relationship between kilograms
   and pounds.

7. If a person weighs 110 pounds, how many kilograms does she
   weigh?

In Exercises 8 and 9, sketch a graph on the same coordinate plane.

8. $y = x + 1$

9. $y = x - 1$

10. Tell whether the lines for Exercises 8 and 9 are parallel. Explain.

1._____

2._____

3._____

4._____

5.

6._____

7._____

8.

9.

10.

Name_____

Date _____

1. Sketch a line having the given intercepts.
   x-intercept: 1
   y-intercept: 3

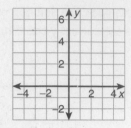

1.

In Exercises 2 and 3, use the equation $y = 3x - 3$.

2. Sketch a quick graph of the line.

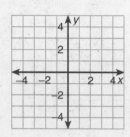

2.

3. Create a table of values and compare the values with the points on the line.

x						
y						

3.

In Exercises 4 and 5, find the slope of the line.

4.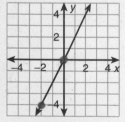

4._____

5.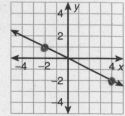

5._____

In Exercises 1–3, use the equation $M = 29.6N$, which relates milliliters M to ounces N. (13.1)

1. A bottle holds 473 milliliters. What is its capacity in ounces? Round your answer to the nearest whole number.

1._____

2. A jar holds 12 ounces. What is its capacity in milliliters? Round your answer to the nearest whole number.

2._____

3. Which is more, a liter (1000 milliliters) or a quart (32 ounces)? Explain.

3.

In Exercises 4–7, use the equation $2x + 3y = 10$. (13.2, 13.3)

4. Decide whether $(3, 2)$ is a solution.

4._____

5. Complete a table of values for the equation.

5.

6. Sketch the graph of the equation.

6.

7. Identify the x- and y-intercepts for the equation.

7._____

In Exercises 8 and 9, find the slope of the line that passes through the points. (13.4)

8. $(3, 2), (-2, -1)$

8._____

9. $(6, 2), (-1, 3)$

9._____

10. Sketch the graphs of $y = 2x + 3$ and $y = x + 3$ on the same coordinate plane. (13.3)

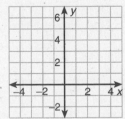

10.

11. Are the lines parallel? Explain. (13.4)

11.

In Exercises 1–3, use the equation F = 3.28M, which relates feet F to meters M. (13.1)

1. A door is 2.5 meters high. What is its height in feet? Round your answer to the nearest whole number.

2. A room is 16 feet long. What is its length in meters? Round your answer to the nearest whole number.

3. Which is a greater distance, a kilometer (1000 meters) or a mile (5280 feet)? Explain.

1._____

2._____

3.

In Exercises 4–7, use the equation 3x + y = 12. (13.2, 13.3)

4. Decide whether (2, 6) is a solution.

5. Complete a table of values for the equation.

6. Sketch the graph of the equation.

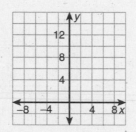

7. Identify the x- and y-intercepts for the equation.

4._____

5.

6.

7._____

In Exercises 8 and 9, find the slope of the line that passes through the points. (13.4)

8. (4, 2), (6, 5)

9. (−2, −3), (−4, −1)

8._____

9._____

10. Sketch the graphs of $y = x + 4$ and $y = 2x + 4$ on the same coordinate plane. (13.3)

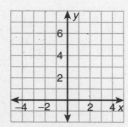

10.

11. Are the lines parallel? Explain. (13.4)

11.

13.6 Short Quiz

Name_____

Date _____

In Exercises 1 and 2, find the slope and *y*-intercept of the line. Then sketch a quick graph.

1. $y = 2x + 5$

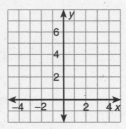

1._____

2. $x - 3y = 6$

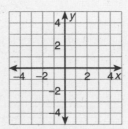

2._____

3. Write an equation of a line with a slope of $-\frac{1}{2}$ and a *y*-intercept of 5.

3._____

4. Graph the equation that you wrote in Exercise 3.

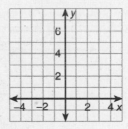

4.

In Exercises 5–7, you are buying beads for a necklace. Small beads cost $0.20 each and large beads cost $0.50. You have $3.00 to spend.

5. Write an algebraic model to represent the necklace beads.

5._____

6. Find the *x*- and *y*-intercepts and graph the model.

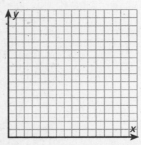

6._____

7. Explain why you could not buy 10 small beads and 3 large beads.

7.

13.8 Short Quiz

Name_____

Date _____

In Exercises 1–4, use the inequality $5x + 4y \geq 20$.

1. Is $(5, 0)$ a solution? Explain.

   1.

2. Is $(0, 5)$ a solution? Explain.

   2.

3. List three other solutions for the inequality.

   3._____

4. Graph the inequality.

   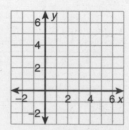

   4.

5. Estimate the distance between the two points. Then use the Distance Formula to check your estimate.

   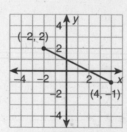

   5._____

6. Estimate the midpoint of the two points. Then use the Midpoint Formula to check your estimate.

   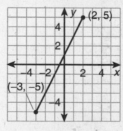

   6._____

7. Find the perimeter of the polygon. Round your result to two decimal places.

   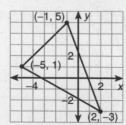

   7._____

In Exercises 1–4, decide whether the ordered pair is a solution of
$4x + y = 20$. **(13.1)**

1. $(4, 4)$ 1._____

2. $(5, 5)$ 2._____

3. $(3, 6)$ 3._____

4. $(6, -4)$ 4._____

In Exercises 5–8, use the equation $2x - y = 8$. **(13.2–13.4)**

5. Create a table of values for the equation.   5.

6. Name the intercepts of the equation.   6._____

7. Use the intercepts to make a quick
   graph of the line.   7.

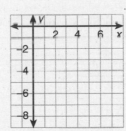

8. Use the graph to find the slope of the line.   8._____

In Exercises 9 and 10, find the slope of the line through the points. (13.4)

9. $(-2, 8)$ and $(3, -2)$ 9._____

10. $(8, 4)$ and $(1, 3)$ 10._____

In Exercises 11–13, decide whether the line is horizontal, vertical, or slanted.
(13.2)

11. $y = -3$ 11._____

12. $x = 4.5$ 12._____

13. $y = 7x$ 13._____

14. Sketch the line that has an *x*-intercept of 3 and a *y*-intercept of −5. (13.3)

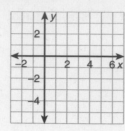

14._____

In Exercises 15 and 16, name the *x*- and *y*-intercepts. Then sketch a quick graph of the line. (13.3)

15. $x + y = 5$

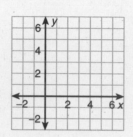

15._____

16. $2x + y = 4$

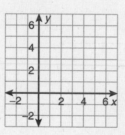

16._____

17. Write an equation of the line. (13.5)

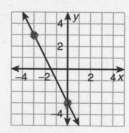

17._____

In Exercises 18 and 19, sketch a graph of the inequality. (13.7)

18. $y \leq -2x$

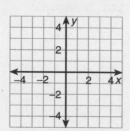

18._____

Passport to Algebra and Geometry

19. $y > 3x + 4$

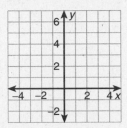

19._____

In Exercises 20 and 21, find the midpoint and length of the line segment. (13.8)

20.

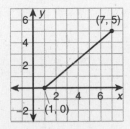

20._____

21.

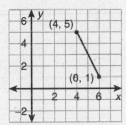

21._____

In Exercises 22–24, use the following:

You are selling pizza for a fund raiser. Slices of plain pizza sell for $1.00 a slice. Pepperoni pizza sells for $2 a slice. At the end of the evening, you find that you have collected a total of $30. (13.6)

22. Write an algebraic model that describes the sale of plain and pepperoni pizza. Let x represent the number of plain pizza slices and y represent the number of pepperoni pizza slices.

22._____

23. You sold 10 slices of plain pizza. How many slices of pepperoni pizza did you sell?

23._____

24. Graph the linear model. List three other solutions.

24.

In Exercises 1–7, use the equation $y = 3x - 5$.

1. Which ordered pair is a solution of the equation? (13.1)   1._____

   a. $(3, 14)$ **b.** $(-2, 11)$ **c.** $(-1, -8)$ **d.** $(0, 5)$

2. What is the *y*-intercept? (13.3)   2._____

   a. 3   **b.** -5 **c.** $\frac{3}{5}$ **d.** $\frac{5}{3}$

3. What is the *x*-intercept? (13.3)   3._____

   a. 3   **b.** 5   **c.** $\frac{3}{5}$ **d.** $\frac{5}{3}$

4. What is the slope? (13.5)   4._____

   a. $\frac{3}{5}$ **b.** $-\frac{3}{5}$ **c.** 3   **d.** $-\frac{3}{2}$

5. What is the value of *y* when the value of *x* is -3? (13.1)   5._____

   a. -14 **b.** -11 **c.** 14   **d.** 4

6. What is the value of *x* when the value of *y* is 1? (13.1)   6._____

   a. 2   **b.** -2 **c.** 0   **d.** -8

7. Which is the graph of the equation? (13.5)   7._____

   a. 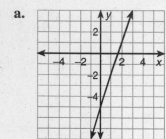   **b.**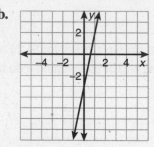

   c. 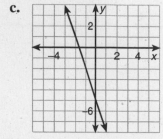   **d.**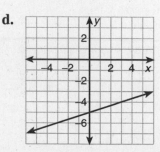

Passport to Algebra and Geometry

In Exercises 8 and 9, what is the slope of the line between the two points? (13.4)

8. $(-2, 3), (4, -6)$

   a. $-\frac{2}{3}$ b. $-\frac{3}{2}$ c. $\frac{3}{2}$ d. $\frac{2}{3}$

   8._____

9. $(-4, -6), (-1, 3)$

   a. 3 b. -1 c. 1 d. $\frac{5}{9}$

   9._____

In Exercises 10 and 11, what is the x-intercept? (13.3)

10. $x - 3y = 6$

    a. 6 b. -3 c. 1 d. -2

    10._____

11. $y = \frac{2}{3}x - 8$

    a. $\frac{2}{3}$ b. -8 c. 12 d. 1

    11._____

In Exercises 12 and 13, what is the y-intercept? (13.3)

12. $4x + 5y = 20$

    a. 4 b. 20 c. 5 d. -5

    12._____

13. $y = \frac{1}{2}x - 1$

    a. 1 b. $\frac{1}{2}$ c. 2 d. -1

    13._____

In Exercises 14–16, what is the equation of the line? (13.5)

14. Slope $= 5$, y-intercept $= -3$

    a. $5x - 3y = -15$ b. $y = 5x - 3$
    c. $y = -3x + 5$ d. $-3x + 5y = 0$

    14._____

15. Passes through points $(3, -2)$ and $(-2, 8)$; y-intercept $= 4$

    a. $y = 3x - 2$ b. $y = -2x + 8$
    c. $y = -2x + 4$ d. $y = 2x + 4$

    15._____

16. x-intercept $= 1$, y-intercept $= 3$

    a. $y = x + 3$ b. $y = \frac{1}{3}x + 1$
    c. $y = -3x + 3$ d. $y = -3x + 1$

    16._____

In Exercises 17–19, which equation describes the graph? (13.7)

17.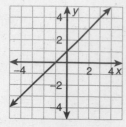

a. $y = x + 1$

b. $y = 2x$

c. $y = 2x + 1$

d. $y = 2x + 3$

17._____

18.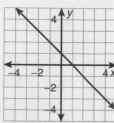

a. $y = x$

b. $y = x + 1$

c. $y = x - 1$

d. $y = -x + 1$

18._____

19.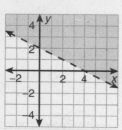

a. $y < -\frac{1}{2}x - 2$

b. $y \leq \frac{1}{2}x - 2$

c. $y \geq -\frac{1}{2}x + 2$

d. $y > -\frac{1}{2}x + 2$

19._____

20. What is the midpoint of the two points? (13.8)

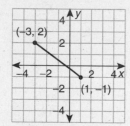

a. $\left(1, \frac{1}{2}\right)$

b. $\left(1, -\frac{1}{2}\right)$

c. $\left(-1, -\frac{1}{2}\right)$

d. $\left(-1, \frac{1}{2}\right)$

20._____

In Exercises 21 and 22, consider the following:

You are buying exactly 20 yards of ribbon for a wreath. You can buy ribbon in 2-yard packages or in 5-yard bolts. How many packages or bolts of ribbon could you buy? (13.6)

21. Which equation represents the situation algebraically?

a. $20 = 2x + 5$

b. $2x + 5y = 20$

c. $5y = 2x + 20$

d. $5 + 2 = 20y$

21._____

22. Which would NOT be a solution of the equation?

a. $(5, 2)$ b. $(2, 5)$ c. $(10, 0)$ d. $(0, 4)$

22._____

In Exercises 1–5, consider the equation $2x + 3y = 12$.

1. Create a table of values for the equation. (13.1)

    1.

2. Name the intercepts of the equation. (13.3)

    2. _____

3. Use the intercepts to sketch a graph of the line. (13.3)

    3.

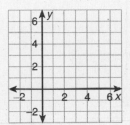

4. Use the graph to find the slope of the equation. (13.4)

    4. _____

5. Write the equation in slope-intercept form. (13.5)

    5. _____

6. Find the slope of the line that passes through $(3, -5)$ and $(-3, 8)$. (13.4)

    6. _____

In Exercises 7–10, use the equation $3x + 6y = 18$.

7. Find the slope and y-intercept of the equation. (13.5)

    7. _____

8. Name three points that line on the line. (13.1)

    8. _____

9. Name three points that would not lie on the line. (13.1)

    9. _____

10. Sketch a graph of the line. (13.5)

    10.

In Exercises 11–13, sketch a line to fit the description. Then write its equation.

11. Horizontal (13.1)

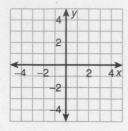

11._____

12. Vertical (13.1)

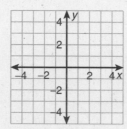

12._____

13. Slanted (13.5)

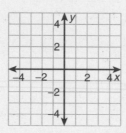

13._____

14. The x-intercept of a line is 10 and the y-intercept is $-\frac{5}{2}$. Write the equation of the line. (13.5)

14._____

In Exercises 15 and 16, identify the points. Then find their midpoint and calculate the distance between them. (13.8)

15.

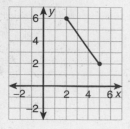

15._____

16.

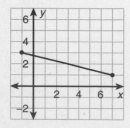

16._____

In Exercises 17 and 18, sketch a graph of the inequality. (13.7)

17. $y > \frac{5}{2}x + 4$

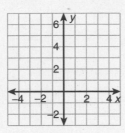

17. _____

18. $y \leq -\frac{2}{3}x - 1$

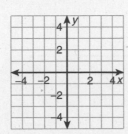

18. _____

19. The enrollment of Cabin John Middle School is shown in the table. Use a scatter plot to estimate the enrollment for 1998. (13.6)

19. _____

Year	1993	1994	1995	1996	1997	1998
Enrollment	275	301	320	350	390	???

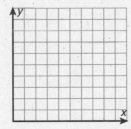

In Exercises 20–22, you are saving up for a game system. You earn $3.00 per hour babysitting and $5.00 an hour mowing lawns. You need at least $35.00 for the game system. (13.7)

20. Write an algebraic model that represents the situation.

20. _____

21. Sketch a graph of the algebraic model.

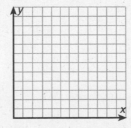

21. _____

22. Identify one point that is part of the solution and one point that is not part of the solution.

22. _____

Name_____

Date _____

In Exercises 1 and 2, find the mean, median, and mode of the data.

1. 11, 13, 25, 16, 12, 14, 12, 23, 10, 19, 22, 12

1._____

2. 4, 2.75, 4, 3, 5.5, 4.75

2._____

In Exercises 3–5, use the line plot.

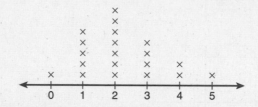

3. List the numbers on the line plot in increasing order.

3._____

4. Find the mean, median and mode of the numbers.

4._____

5. Describe a real-life situation that could be represented by the line plot. Explain your thinking.

5.

6. Write a set of seven numbers for which the mean is 5, the mode is 7, and the median is 6.

6.

In Exercises 7 and 8, use the set of data, showing the average high temperatures in Rockville for June:

78°, 77°, 78°, 79°, 81°, 83°, 82°, 79°, 67°, 68°, 72°, 76°, 79°, 84°, 87°, 87°, 87°, 89°, 91°, 93°, 89°, 91°, 86°, 82°, 83°, 81°, 78°, 81°, 86°, 85°.

7. Order the data in a stem-and-leaf plot.

7._____

8. Draw a histogram to represent the data.

8.

Passport to Algebra and Geometry

Name_____

Date _____

In Exercises 1–4, use the box and whisker plot. There are 20 numbers in the collection and each number is different.

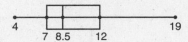

4   7 8.5   12   19

1. Name the smallest and largest numbers.   1._____

2. Name the first, second, and third quartiles.   2._____

3. What percent of the numbers are greater than 12?   3._____

4. What percent of the numbers are between 7 and 8.5?   4._____

In Exercises 5 and 6, find the sum and difference of the matrices.

5. $\begin{bmatrix} 2 & -4 \\ -5 & 9 \end{bmatrix} \begin{bmatrix} -14 & 12 \\ -4 & -6 \end{bmatrix}$ 5.

6. $\begin{bmatrix} -4 & 6 & 0 \\ 3 & -2 & 5 \end{bmatrix} \begin{bmatrix} 2 & 5 & -7 \\ -3 & 0 & 7 \end{bmatrix}$ 6.

7. Find two matrices whose sum is the given matrix. (Zero may not be used.)   7.

$\begin{bmatrix} 1 & 2 \\ 3 & 4 \end{bmatrix}$

Name_____

Date _____

In Exercises 1–10, use the data.

Boys' Math Test Scores: 89, 78, 66, 79, 86, 49, 95, 87, 57, 85, 98, 80

Girls' Math Test Scores: 76, 97, 81, 52, 90, 76, 80, 99, 89, 82, 78

1. Create a double stem-and-leaf plot for the data. (14.2)

1._____

2. Find the median of test scores for the boys. (14.1)

2._____

3. Find the mean of test scores for the girls. (14.1)

3._____

4. Find the mode of the test scores for the girls. (14.1)

4._____

5. Use the stem-and-leaf plot to order all of the test scores from least to greatest. (14.2)

5.

6. Find the first quartile. (14.3)

6._____

7. Find the second quartile. (14.3)

7._____

8. Find the third quartile. (14.3)

8._____

9. Draw a box-and-whisker plot of the data. (14.3)

9.

10. What does the box-and-whisker plot tell you about the test scores? (14.3)

10.

11. Find the sum and difference of the matrices. (14.4)

11.

$$\begin{bmatrix} -4 & 7 & -1 \\ 3 & -8 & 0 \end{bmatrix}, \begin{bmatrix} 8 & -2 & -5 \\ 9 & -3 & -11 \end{bmatrix}$$

In Exercises 1–8, use the data showing ticket sales of two roller coasters during ten consecutive hours.

Small Roller Coaster Ticket Sales: 85, 87, 91, 76, 59, 63, 77, 82, 98, 85
Large Roller Coaster Ticket Sales: 49, 63, 78, 89, 94, 99, 92, 91, 87, 76

1. Create a double stem-and-leaf plot for the data. (14.2)

1.

2. Find the median of ticket sales for the small roller coaster. (14.1)

2._____

3. Find the mean of ticket sales for the larger roller coaster. (14.1)

3._____

4. Find the mode of ticket sales for the small roller coaster. (14.1)

4._____

5. Use the stem-and-leaf plot to order all of the ticket sales from least to greatest. (14.2)

5._____

6. Find the first quartile. (14.3)

6._____

7. Find the second quartile. (14.3)

7._____

8. Find the third quartile. (14.3)

8._____

9. Draw a box-and-whisker plot of the data. (14.3)

9.

10. What does the box-and-whisker plot tell you about ticket sales for roller coasters? (14.3)

10.

11. Find the sum and difference of the matrices. (14.4)

11.

$$\begin{bmatrix} -2 & -4 & -3 \\ 0 & 5 & -6 \end{bmatrix}, \begin{bmatrix} -8 & 5 & 3 \\ 10 & 7 & -2 \end{bmatrix}$$

14.6 Short Quiz

Name _____

Date _____

In Exercises 1–3, write a polynomial for the description given. Then list its terms.

1. Monomial

   1. _____

2. Binomial

   2. _____

3. Trinomial

   3. _____

4. Draw algebra tiles to show the polynomial $3x^2 + 2x + 6$.

   4.

In Exercises 5 and 6, simplify the polynomial and write it in standard form.

5. $4m + 5m^2 - 3 + 2m - 4m^2$

   5. _____

6. $3.5x^3 - 2x^2 + 0.5x^3 + 2x^2 + 4x$

   6. _____

In Exercises 7 and 8, perform the operations. Use a horizontal format.

7. $(3x^2 + 4x + 7) + (2x^2 - 9x - 2)$

   7. _____

8. $(5y^2 - 3y + 12) - (6y^2 + 2y + 5)$

   8. _____

In Exercises 9 and 10, perform the operations. Use a vertical format.

9. $\quad 4b^3 - 3b^2 + 5b + 6$
   $\quad + 6b^3 - 2b^2 - 7b + 9$

   9. _____

10. $\quad 7c^3 - 4c^2 - 3c + 1$
    $\quad - (4c^3 + 5c^2 - c + 7)$

    10. _____

Name_____

Date _____

In Exercises 1–3, find the product.

1. $3d(4d^2 + 2d - 1)$

    1._____

2. $-5f^2(2f^2 - 4f + 5)$

    2._____

3. $w^2(-5w^3 + 4w^2 - 3w + 2)$

    3._____

In Exercises 4–7, use the rectangular prism.

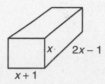

4. Write an expression for the area of the top of the prism.

    4._____

5. Write an expression for the surface area of the prism.

    5._____

6. Write an expression for the volume of the prism.

    6._____

7. Assign a value between 1 and 5 for x. Find the volume of the prism using your value for x.

    7._____

In Exercises 8 and 9, find the product using the Distributive Property.

8. $(x + 5)(2x + 6)$

    8._____

9. $(3x + 7)(4x + 8)$

    9._____

In Exercises 10 and 11, use the triangle.

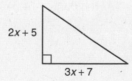

10. Write an expression for the area of the figure.

    10._____

11. Find the area if $x = 5$.

    11._____

In Exercises 1–7, use the table, which lists the top ten medal winning countries in the 1994 Winter Olympics. *(Source: 1997 World Almanac)*

Country	Total Medals	Country	Total Medals
Norway	26	Canada	13
Germany	24	Switzerland	9
Russia	23	Austria	9
Italy	20	South Korea	6
United States	13	Finland	6

1. Find the mean medal total. (14.1)

1._____

2. Find the median medal total. (14.1)

2._____

3. Find the mode of the medal totals. (14.1)

3._____

4. Which measure of central tendency do you think best represents the data? Explain. (14.1)

4.

5. Make a stem-and-leaf plot to organize the data. Let the stems represent the tens digits and the leaves represent the units digits. (14.2)

5.

6. List the first, second, and third quartiles for the data. (14.3)

6._____

7. Use the results of the stem-and-leaf plot to draw a histogram for the data. (14.1)

7.

Passport to Algebra and Geometry

In Exercises 8–11, use the double
stem-and-leaf plot. (14.2)

Group 1		Group 2
	3	5
9 2 0	2	0 0 2 3 6
8 7 5	1	3 7
9 8 3 2 1 1	0	2 2 3 9

5 | 1 | 3 represents 15 and 13.

8. Name the mean for each group. (14.1)

8. _____

9. Identify the first, second, and third quartile for Group 1. (14.3)

9. _____

10. Identify the first, second, and third quartile for Group 2. (14.3)

10. _____

11. Construct a box-and-whisker plot for the data of one group. Identify
the group you use. (14 3)

11. _____

In Exercises 12 and 13, find the first, second, and third quartile for the data.
(14.3)

12. 8, 5, 4, 3, 9, 2, 6, 4, 1, 2

12. _____

13. 28, 36, 49, 27, 62, 43, 54, 31, 27, 69, 40, 31

13. _____

In Exercises 14 and 15, find the sum or difference. (14.4)

14. $\begin{bmatrix} 3 & -5 \\ -4 & 9 \end{bmatrix} + \begin{bmatrix} 4 & 7 \\ 2 & -8 \end{bmatrix}$

14. _____

15. $\begin{bmatrix} 5 & 7 & -8 \\ 3 & -6 & -1 \\ -1 & 3 & -2 \end{bmatrix} - \begin{bmatrix} 3 & -8 & -2 \\ 2 & -6 & -1 \\ -4 & 5 & 6 \end{bmatrix}$

15. _____

In Exercises 16 and 17, simplify the polynomial and write it in standard form. Then state whether the result is a monomial, binomial, or trinomial. (14.5)

16. $-4x^2 + 3x - 5x^2 + 6x + x^2$

16._____

17. $7 + 2x^2 + 8x - 3 - 6x^2 - 1$

17._____

In Exercises 18 and 19, add or subtract the polynomials and simplify. (14.6)

18. $(3x^2 + 5x - 7) + (-2x^2 - 5x + 7)$

18._____

19. $(4x^3 - 5x^2 + 6x - 7)$
    $- (3x^3 + 2x^2 - 8x + 9)$

19._____

In Exercises 20–22, multiply the polynomials. (14.7)

20. $4(3k^2 + 7)$

20._____

21. $5x^2(8x^2 + 9x - 4)$

21._____

22. $(2n + 1)(9n + 5)$

22._____

In Exercises 23–25, use the figure.
(14.7, 14.8)

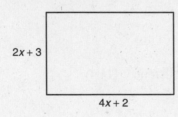

2x + 3

4x + 2

23. Write an expression for the perimeter of the figure.

23._____

24. Write an expression for the area of the figure.

24._____

25. Choose a value for x between 1 and 5. Find the perimeter and area of the figure using your value for x.

25._____

Passport to Algebra and Geometry

In Exercises 1 and 2, what is the mean of the numbers? (14.1)

1. 24, 29, 28, 25, 29

   a. 29   **b.** 25   **c.** 27   **d.** 28

1._____

2. 2.63, 9.61, 3.057, 8.39, 5.12, 5.76

   a. 5.76   **b.** 5.12   **c.** 3.057   **d.** 6.003

2._____

In Exercises 3 and 4, what is the median of the numbers? (14.1)

3. 70, 72, 75, 75, 78, 80, 80

   a. 75   **b.** 80   **c.** 78   **d.** 72

3._____

4. 9.1, 11.5, 8.4, 7.0, 9.3, 6.1, 10.9

   a. 8.4   **b.** 7.0   **c.** 8.9   **d.** 9.1

4._____

5. What is the mode of the listed data: 95.6, 65.9, 76.5, 96.8, 76.5, 57.6? (14.1)

   a. 76.4   **b.** 76.5   **c.** 65.9   **d.** 95.6

5._____

In Exercises 6–8, use the box-and-whisker plot, showing the ages of media specialists in the Spring City Schools. (14.3)

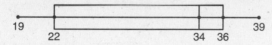

19   22   34   36   39

6. What is the median?

   a. 19   **b.** 22   **c.** 34   **d.** 36

6._____

7. Which number is NOT a quartile?

   a. 19   **b.** 22   **c.** 34   **d.** 36

7._____

8. About what percentage of the media specialists are between the ages of 34 and 39?

   a. 25%   **b.** 35%   **c.** 50%   **d.** 65%

8._____

In Exercises 9–12, use the double stem-and-leaf plot, showing scores of the winning and losing Super Bowl teams from 1987 through 1996. (14.2)

(Source: 1997 World Almanac)

Losers		Winners
	5	2 5
	4	2 9
	3	0 7 9
6 4 0	2	0 0 7
9 7 7 6 3 0 0	1	

0|2|0 represents 20 and 20.

9. What is the mean for the scores of the losing teams?

   a. 7   **b.** 20

   c. 17.2   **d.** 4.2

   9._____

10. What is the mean for the scores of the winning teams?

    a. 20   **b.** 9   **c.** 4.1   **d.** 37.1

    10._____

11. What are the modes, if there are any, for the losing team scores?

    a. 10 and 20   **b.** 10 and 17   **c.** 16 and 26   **d.** None

    11._____

12. What is the mode(s), if there is any, for the winning team scores?

    a. 20   **b.** 20 and 30   **c.** 27 and 37   **d.** None

    12._____

In Exercises 13 and 14, what is the sum or difference of the matrices? (14.4)

13. $\begin{bmatrix} -5 & 2 \\ 6 & -4 \end{bmatrix} + \begin{bmatrix} -8 & 7 \\ 5 & -1 \end{bmatrix}$

    13._____

    a. $\begin{bmatrix} 13 & 9 \\ 11 & -5 \end{bmatrix}$ **b.** $\begin{bmatrix} -13 & 9 \\ 11 & -5 \end{bmatrix}$

    c. $\begin{bmatrix} -13 & 9 \\ 11 & -3 \end{bmatrix}$ **d.** $\begin{bmatrix} -13 & 9 \\ 11 & 5 \end{bmatrix}$

14. $\begin{bmatrix} -3 & 6 & 7 \\ 5 & 0 & -4 \end{bmatrix} - \begin{bmatrix} -5 & -2 & 1 \\ -5 & 4 & -4 \end{bmatrix}$

    14._____

    a. $\begin{bmatrix} -8 & 4 & 8 \\ 0 & 4 & -8 \end{bmatrix}$ **b.** $\begin{bmatrix} 2 & 8 & 6 \\ 10 & -4 & 0 \end{bmatrix}$

    c. $\begin{bmatrix} 2 & 8 & 6 \\ 0 & 4 & -8 \end{bmatrix}$ **d.** $\begin{bmatrix} -8 & 4 & 8 \\ 10 & -4 & 0 \end{bmatrix}$

In Exercises 15 and 16, what is the simplified version of the polynomial? (14.5)

15. $3x^2 + 8 - 4x + x^2 - 6x + 2$

    15._____

    a. $4x^2 - 10x + 10$ **b.** $3x^2 - 2x + 6$

    c. $2x^2 - 10x + 10$ **d.** $4x^2 + 10x - 10$

16. $4n^3 + 7n^2 - 5n^2 + 2 - 5 + 6n - n^3$

    16._____

    a. $5n^3 + 2n^2 + 6n - 3$ **b.** $3n^3 - n^2 + 2n + 3$

    c. $3n^3 - 2n^2 + 6n - 3$ **d.** $3n^3 + 2n^2 + 6n - 3$

In Exercises 17 and 18, what is the sum or difference of the polynomials? (14.6)

17. $4x^3 - 6x^2 + 9x - 1$

        $\underline{+\ 2x^3 + 5x^2 - 9x + 7}$

17. _____

    a. $6x^3 - x^2 - x + 6$ **b.** $2x^3 + x^2 + 6$

    c. $6x^3 - x^2 + 6$ **d.** $2x^3 - 11x^2 + 18x - 8$

18. $(3x^3 - 5x^2 + 10x - 1) - (4x^3 + 6x^2 + 5x - 1)$

18. _____

    a. $-x^3 + x^2 - 5x - 2$ **b.** $7x^3 + x^2 + 15x - 2$

    c. $-x^3 - 11x^2 - 5x$ **d.** $7x^3 - 11x^2 - 5x$

In Exercises 19–21, what is the product of the polynomials? (14.7, 14.8)

19. $5(3w - 4)$

19. _____

    a. -5 **b.** $15w - 20$ **c.** $15w + 20$ **d.** $-15w + 20$

20. $-2t^2(3t^2 - 4t + 7)$

20. _____

    a. $6t^4 + 4t^3 - 14t^2$ **b.** $-6t^2 + 8t - 14$

    c. $-6t^4 + 8t^3 - 14t^2$ **d.** $6t^4 - 8t^3 - 14t^2$

21. $(3x + 4)(x + 5)$

21. _____

    a. $3x^2 + 12x + 20$ **b.** $3x^2 + 19x + 20$

    c. $4x^2 + 15x + 20$ **d.** $4x^2 + 12x + 5$

In Exercises 22 and 23, use the figure. (14.7, 14.8)

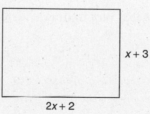

$x + 3$

$2x + 2$

22. Which expression describes the
perimeter of the figure?

22. _____

    a. $(x + 3)(2x + 2)$

    b. $(x + 3) + (2x + 2)$

    c. $(x + 3) - (2x + 2)$

    d. $2(x + 3) + 2(2x + 2)$

23. Which expression describes the area of the figure?

23. _____

    a. $(x + 3)(2x + 2)$ **b.** $(x + 3) + (2x + 2)$

    c. $(x + 3) - (2x + 2)$ **d.** $2(x + 3) + 2(2x + 2)$

In Exercises 1–7, use the table, which shows the All-Time Top 10 Movies through March, 1996. *(Source: 1997 World Almanac)*

Movie Title	Gross (in millions)	Movie Title	Gross (in millions)
E.T.: The Extra-Terrestrial	$399.8	*Home Alone*	285.8
Jurassic Park	357.1	*Return of the Jedi*	263.7
Forrest Gump	329.7	*Jaws*	260.0
Star Wars	322.7	*Batman*	251.2
The Lion King	312.9	*Raiders of the Lost Ark*	242.4

1. Find the mean gross revenue. (14.1)

1. _____

2. Find the median gross revenue. (14.1)

2. _____

3. Find the mode of the gross revenues. (14.1)

3. _____

4. Which measure of central tendency do you think best represents the data? Explain. (14.1)

4.

5. Make a stem-and-leaf plot to organize the data. Label the stems and leaves to show what they represent. (14.2)

5.

6. Use the results of the stem-and-leaf plot to draw a histogram for the data. (14.1, 14.2)

6.

7. Draw a box-and-whisker plot for the data in the table. (14.3)

7.

In Exercises 8 and 9, use {80, 52, 65, 76, 75, 97, 69, 86, 99, 80, 46, 82, 79, 81, 68}. (14.3)

8. List the first, second and third quartiles for the data.

8._____

9. Construct a box-and-whisker plot for the data.

9._____

In Exercises 10–14, use {1, 4, 2, 1, 2, 9, 7, 3, 4, 8}. (14.1)

10. Find the mean.

10._____

11. What data could be added so that the mean is 4? Explain.

11._____

12. Find the median of the given data.

12._____

13. What data could be added so that the median is 4? Explain.

13._____

14. What data could be added so that the mode is 4?

14._____

In Exercises 15 and 16, use this data.

Science Grades, First Semester: 64, 72, 78, 70, 81, 85, 87, 91, 83, 86
Science Grades, Second Semester: 76, 82, 91, 95, 88, 90, 78, 89, 93, 95

15. Make a double stem-and-leaf plot to organize the data. (14.2)

15.

16. What can you determine from analyzing the data? Explain. (14.2)

16.

In Exercises 17 and 18, find the sum or difference. (14.4)

17. $\begin{bmatrix} -4 & 5 & 0 \\ 10 & 0 & -3 \end{bmatrix} + \begin{bmatrix} 3 & -5 & 7 \\ -4 & 6 & -9 \end{bmatrix}$

17. _____

18. $\begin{bmatrix} -1 & 3 & -7 \\ 0 & 4 & -3 \\ 5 & 1 & 6 \end{bmatrix} - \begin{bmatrix} -8 & 2 & 1 \\ -7 & 9 & -3 \\ -5 & -1 & 6 \end{bmatrix}$

18. _____

19. Create a pair of matrices whose sum is $\begin{bmatrix} 4 & -5 \\ -2 & 0 \end{bmatrix}$. (14.4)

19. _____

In Exercises 20 and 21, simplify the polynomials and write in standard form. Then describe it as a monomial, binomial, or trinomial. (14.5)

20. $4 + 3x - 7x + 4x - 3x^2 - 4$

20._____

21. $10x^2 - 4x + 6x - x^2 + 5 - 9x^2 - x$

21._____

22. Write a trinomial. Identify its terms. (14.5)

22._____

In Exercises 23 and 24, add or subtract the polynomials and simplify. (14.6)

23. $\quad 4x^4 + 6x^3 - 5x^2 - 7x + 1$
    $\quad \underline{+ \ 2x^4 - 3x^3 \qquad\quad + 7x - 5}$

23._____

24. $(-8x^4 + 4x^3 - x + 2) - (9x^4 - 8x^2 + 7x + 4)$

24._____

In Exercises 25 and 26, multiply the polynomials. (14.7, 14.8)

25. $8x^2(4x^3 - 3x + 7)$

25._____

26. $(4x + 5)(4x + 5)$

26._____

Answers to Formal Assessment

■ 1.2 Short Quiz

1. Multiples of 6; 30, 36, 42
2. Decreasing by 6; 30, 24, 18
3. The first half of the race is faster, the time is shorter (less).
4. The sum of thirty-five and four hundred seventy-nine equals five hundred fourteen.
5. 547.6   6. 65.57   7. 27
8. Patterns and descriptions will vary. (Description should match pattern.)

■ 1.4 Short Quiz

1. Eight raised to the third power (or 8 cubed) equals five hundred twelve.
2. The square root of 1.44 equals 1.2.
3. 4^5; 1024   4. 1.01   5. 50
6. $8 \cdot (2 + 3) \div 4 = 10$ 7. $27 \div (42 - 39)$; 9
8. Possible answer: $9 \cdot (8 - 6) \div (2 + 4) = 3$; $(9 \cdot 8) - (6 \div 2) + 4 = 73$; $3 < 73$

■ Mid-Chapter Test, 1-A

1. Multiples of 4; 16, 20, 24
2. Fractions with denominators increasing by 1; $\frac{1}{5}, \frac{1}{6}, \frac{1}{7}$
3. 6 raised to the 3rd power; 216
4. The quotient of 203 and 29; 7
5. 32   6. 6
7. Possible answer: 3 water rides, 3 ground rides, total 48 tickets
8. 16 inches
9. Possible answer: 2 - 8 × 8's; 8 - 4 × 4's

■ Mid-Chapter Test, 1-B

1. Multiples of 7; 28, 35, 42
2. Numbers increasing by $\frac{1}{2}$; 2, $\frac{5}{2}$, 3
3. 4 raised to the 4th power; 256
4. The quotient of 243 and 27; 9
5. 30   6. 16
7. Possible answer: 5 Interactive, 5 Seated; total 50 tokens
8. 12 feet   9. 2 - 6 × 6's; 8 - 3 × 3's

■ 1.6 Short Quiz

1. 40   2. 18   3. $n + 3$
4. The quotient of t and 8   5. 7
6. 45   7. Yearbook, computer
8. Possible answer: Literary and Yearbook; they are both small and could be related.
9. Possible answer: The estimated membership if Literary and Yearbook are combined is 57.

■ 1.8 Short Quiz

1. Check students' drawings.
2.   3. 42 ft²   4. Decagon

   20 diagonals
5. Not a polygon, one side is curved.
6. 120; 143; 168; 195; 224; the sum increases by 2 with each subsequent product (+23, +25, +27, +29)
7. Answers will vary.

■ Chapter Test, 1-A

1. Multiples of 8; 40, 48, 56
2. Decreasing by 11; 64, 53, 42
3. 58.917   4. 416   5. 31   6. 10
7. 4^5; 1024   8. $\left(\frac{1}{2}\right)^6$; $\frac{1}{64}$ 9. 3.5
10. 20.25   11. 23 in.   12. 15   13. 9
14. 45   15. 9   16. About 34
17. Toronto and St. Louis; Detroit
18. Answers will vary.   19. Not a polygon
20. Decagon   21. Pentagon
22. Lincoln Park
23. Audubon, Cleveland, Dallas
24. 15 min   25. 3 mi   26. 20 min

■ Chapter Test, 1-B

1. a   2. b   3. c   4. a   5. d   6. c
7. d   8. a   9. a   10. c   11. a   12. b
13. c   14. d   15. a or c   16. c   17. d
18. b   19. a   20. b   21. c   22. c   23. d

■ **Chapter Test, 1-C**

1. Multiply each term by 4 to find the succeeding term; 512, 2048, 8192

2. Each term is decreased by one more than the preceding term; 90, 85, 79

3. 1, 2, 3, 5, 8, 13

4. 1.4, 2, 2.45, 2.83, 3.16, 3.46

5. 0.02263   **6.** 5.2   **7.** 42   **8.** 11.2

9. $(4.5)^4$; 410.0625   **10.** $\left(\frac{4}{5}\right)^4$; $\frac{256}{625}$

11. 26 inches

12.

246	Even Number	Product
246	2	492
246	4	984
246	6	1476
246	8	1968
246	10	2460

Possible answers: The number given by the last two digits decreases by 8 as the number of hundreds increases by 5; each product is $(500 - 8)$ or 492 greater than the previous product.

13. 189   **14.** 6   **15.** 384   **16.** 9

17. ≈ $1700   **18.** Advanced degree

19. One possible answer: Males earn more across the board!

20. Order of operations has been totally ignored; only left-to-right rule is used.

Process should be:

$6 + 4 \cdot (8 - 5)^3 - 20 \div 2$

$6 + 4 \cdot (3)^3 - 20 \div 2$

$6 + 4 \cdot 27 - 20 \div 2$

$6 + 108 - 20 \div 2$

$6 + 108 - 10$

$114 - 10$

104

21. A 5-sided figure with all sides congruent

22. An 8-sided figure; one possible sketch:

23. 50 minutes faster to run than to walk

24. $8(10) + 2(15)$

25. 110 minutes or 1 hour, 50 minutes

26. Answers will vary.

■ **2.2 Short Quiz**

1. $3x + 6$ **2.**

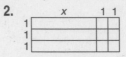

3. $15x + 60$ **4.** $12x + 3$ **5.** $6b + 12$

6. Answers will vary. One possible answer: $3p + 7p + 6$

7. $2x + 4(x + 2)$ or $6x + 8$ **8.** 44

■ **2.4 Short Quiz**

1. $11a + 3$ **2.** $7a^2x + 17ax$

3. $15(w - r)$ or $15w - 15r$

4. Answers will vary. One possible answer: $7x + 3x + 4y$

5. 450   **6.** $9 \cdot n = 72$; $n = 8$

7. $85 - n = 76$; $n = 9$

8. $r = 7$; equations will vary.

9. $e = 15$; equations will vary.

■ **Mid-Chapter Test, 2-A**

1. $8a + 32$ **2.** $5x + 15y + 25$ **3.** $13c + 3$

4. $11x + 14$ **5.** 100   **6.** $2x + 20$ **7.** $10x$

8. $p = 38$ **9.** $v = 35$ **10.** $b = 39$

11. Answers will vary, but all responses should be a triangle with a perimeter of $6x$.

■ **Mid-Chapter Test, 2-B**

1. $7t + 35$ **2.** $3i + 24 + 12j$ **3.** $14u + 15$

4. $7b + 45$ **5.** 308   **6.** $2y + 18$ **7.** $9y$

8. $a = 16$ **9.** $k = 24$ **10.** $y = 37$

11. Answers will vary, but all responses should be a rectangle with a perimeter of $6z$.

■ **Short Quiz 2.6**

1. $y = 13$ **2.** $v = 96$ **3.** $f = 1.6$

4. $w = 4$ units

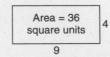

5. $4c = 24$; $c = 6$ **6.** $v \div 6 = 3$; $v = 18$

7. Answers will vary; One possible answer: 2 times the number of fingers, f, equals 10 fingers

8. $3n - 5$ **9.** $7n + 2$

Short Quiz 2.8

1. $t + 6 = 15$; 9   **2.** $w - 43 = 8$; 51

3. A possible verbal sentence: 16 equals the value of c plus 5; 11

4. $398 + n = 500$ or $500 - 398 = n$; 102

5. Total cost $=$ amount saved $+$ amount still needed

6. Respectively; $47.25;
$12.50 + $9.75 + $15.75 or $38;
x (amount needed)

7. $47.25 = $12.50 + $9.75 + $15.75 + x or $47.25 = $38 + x

8. $9.25

Chapter Test, 2-A

1. $6 + 8$ or 14   **2.** $8c + 16$ **3.** $9p$

4. $4x^2 + 5x$ **5.** $6y + 12$ **6.** $7b + 22$

7. $n = 11$ **8.** $q = 7$ **9.** $r = 4$

10. $n + 46$ **11.** $\dfrac{n}{28}$ **12.** $n - 23 = 18$

13. $9n = 63$ **14.** $f = 38$ **15.** $y = 159$

16. $g = 9$ **17.** $a = 48$ **18.** $y \leq 41$

19. $p > 8.5$ **20.** $m \geq 90.5$ **21.** $h \leq 48$

22. $m - $10.50 = 27.75; $m = 38.25

23. $2.50x \geq 15.00; $x \geq 6

24. $215 \div 5 = n$; $n = 43$ minutes

25. $2(L + 5)$ or $2L + 10$

26. $5L$ **27.** 26 units; 40 square units

28. Answers will vary, but values for L should be between 7 and 8.

29. 222 years (from 1999)   **30.** 1960

31. Answers will vary.   **32.** Answers will vary.

33. Answers will vary.

Chapter Test, 2-B

1. c   **2.** d   **3.** a   **4.** b   **5.** d

6. d   **7.** a   **8.** b   **9.** c   **10.** a

11. b   **12.** c   **13.** c   **14.** c   **15.** a

16. a   **17.** b   **18.** c   **19.** c   **20.** a

21. d   **22.** a   **23.** d   **24.** a   **25.** b

Chapter Test, 2-C

1. $40 + 50 + 55$ or 145

2. $3z + fz + jz$ **3.** $3i + 4j + 6k$

4. $c^2 + 3c$ **5.** $5d + 8$ **6.** $x^2 + 19ax$

7. 42   **8.** 43   **9.** $q = 18$ **10.** $r = 5$

11. $5n + 6$ **12.** $59 - n$ **13.** $(n - 6)(n + 6)$

14. $v = 6.03$ **15.** $s = 437$

16. $w = 5$ **17.** $x = 1296$

18. $39.95 + $9.49 = j$ **19.** $0.73 \cdot 10 = d$
$j = 49.44 $d = 7.3$ miles

20. $w + $12.75 = 30.00; $w = 17.25

21. Answers will vary.   **22.** $h \geq 19$

23. $g < 79.2$ **24.** $x \geq 9$ **25.** $e > 25.6$

26. Total miles $=$ distance traveled $+$ distance remaining

27. $1387 = 375 + 438 + x$ **28.** $x = 574$ miles

29. $2(w + w + 3)$ or $2(2w + 3)$ or $4w + 6$

30. $P = 56$ **31.** $w \cdot (w + 3)$ or $w^2 + 3w$

32. $A = 270$

Short Quiz 3.2

1.

2. $<$ **3.** $>$ **4.** -12 **5.** $+17$

6. $2 + (-5) = x$; $x = -3$

7. Answers will vary. Possible values for r and s: -1 and -2; 3 and -6

8. $-2 + 5 = f$ **9.** 3; third floor

Short Quiz 3.4

1. $-5 + 7 + (-3) = -1$

2. $-9 + (-3) + 15 = 3$

3. $2r + 7$; 13   **4.** $7r + 8$; 29

5. $-3 - (-7) = 4$ **6.** $-6 - 9 = -15$

7. $-3s + 8$ **8.** $-4h - 5$

9. Answers will vary. Possible answer: It is $-3°C$ outside, then the temperature rises $8°C$. What is the temperature now?; $5°C$

Mid-Chapter Test, 3-A

1.

2.

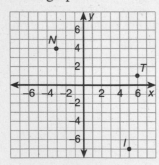

A number line from −4 to 6 with points marked at −4, 0, and 6.

3. +10, +10   **4.** Answers will vary.

5. $u = 15$ **6.** $v = -9$ **7.** 14

8. −2   **9.** −3

10. July and December, she had a change of $30.

11. $+10 + 30 + 20 + 10 + (-20) + 10 + (-30) = c$; $c = +$30$

■ **Mid-Chapter Test, 3-B**

1. −6, 3, 4, 5   **2.** −2, 0, 5, 7   **3.** +8, +8

4. Answers will vary.   **5.** $t = -11$

6. $g = -2$ **7.** −2   **8.** 2   **9.** 2

10. Saturday; a change of 0.08 seconds.

11. Saturday; 34.92

■ **Short Quiz 3.6**

1. $6 \cdot -4 = -24$ **2.** $(-5)(-8)(2) = 80$

3. 20   **4.** $x = -7$ **5.** $m = 7$

6. $c = -18$ **7.** $y = -9$ **8.** −18

9. 51   **10.** 0.5   **11.** Answers will vary.

■ **Short Quiz 3.8**

1. −5   **2.** 25   **3.** −7.7   **4.** −11.8

5. $15 = -10 + m$; $m = 25$

6.–9. See graph.

A coordinate grid with points N in Quadrant 2 at about (−4, 4), T in Quadrant 1 at about (6, 1), and I in Quadrant 4 at about (6, −6).

6. Quadrant 4   **7.** Quadrant 2

8. Quadrant 1

9. Answers will vary, but point should be in Quadrant 3; $(-x, -y)$

■ **Chapter Test, 3-A**

1. −8, −5, 0, 3, 7   **2.** −5, −4, 0, 3, 8

3. −63, 63   **4.** −30   **5.** +45   **6.** −1

7. −6   **8.** 65   **9.** −63   **10.** 17

11. −4   **12.** $5y + 4$; 19   **13.** $4y - 3z$; 6

14. $10z + 5$; 25   **15.** −2   **16.** −5

17. $-6n = -24$; $n = 4$

18. $x - (-5) = 9$; $x = 4$ **19.** $r = -15$

20. $s = -13$ **21.** $t = 7$ **22.** $u = -54$

23. 20°C   **24.** 4°C   **25.** 0.5°C

26. Answers will vary.

27. $P = 325 - (10 + 35 + 50)$ or $P = 325 - 95$; $P = 230

28. $9.20

29.–35. See graph.

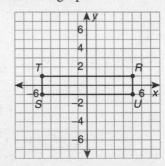

A coordinate grid showing a rectangle with vertices T at (−6, 2), R at (6, 2), S at (−6, −1), and U at (6, −1).

29. Quadrant 1   **30.** Quadrant 3

31. Quadrant 2   **32.** Quadrant 4

33. Rectangle   **34.** 24 units   **35.** 20 sq units

36. No   **37.** Yes   **38.** Yes

■ **Chapter Test, 3-B**

1. d   **2.** c   **3.** a   **4.** d   **5.** b

6. a   **7.** d   **8.** c   **9.** b   **10.** a

11. b   **12.** a   **13.** b   **14.** b   **15.** d

16. a   **17.** c   **18.** c   **19.** a   **20.** b

21. d   **22.** a   **23.** c   **24.** a   **25.** d

■ **Chapter Test, 3-C**

1.

A number line from −5 to 7 with points marked at −3 and 3.

2.

A number line from −5 to 1 with points marked at −5, −4, −2, 0, and 1.

3. Answers will vary.   **4.** Answers will vary.

5. Answers will vary.   **6.** 3   **7.** 306

8. 405   **9.** −25.666 or $-25\frac{2}{3}$ **10.** 14.3

11. $n - 5 = -22$; $n = -17$

12. $n \div -3 = -15$; $n = 45$

13. $-5rs - 2r^2$; 42   **14.** $rs + 4r$; −2

15. $y = -7$ **16.** $g = 25$ **17.** $b = -20$

18. $t = -6.25$

19. $J = \$6.30 + \$8.75 + \$27.90$ or $\$27.90 = J - (\$6.30 + \$8.75)$; $J = \$42.95$

20. $6.20 **21.** $46.55 **22.** 4.25 hr

23.–28. Answers will vary. Possible answer:

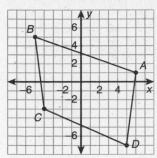

29. Answers will vary. x and y coordinates will be opposites; graph will be a line.

■ **Cumulative Test 1–3**

1. Decreasing by 5 each time. 30, 25, 20

2. Denominator of one fraction becomes numerator of the next, denominators are in sequential order. $\frac{5}{6}, \frac{6}{7}, \frac{7}{8}$

3. Letters of alphabet alternating with corresponding numerical placement of letters. C, 3, D

4. Outermost letters of alphabet, alternating beginning with end, working inward. X, C, W

5. Decreasing by powers of 10. 100, 10, 1

6. Decreasing by one less each time. 66, 60, 55

7. 4,782,969 **8.** 0.26 **9.** 7.42

10. 22.36 **11.** 143.49 **12.** 6.58 **13.** 90

14. 246 **15.** 188 **16.** 3 **17.** -3

18. 7 **19.** 68 **20.** -52 **21.** 35

22. -210 **23.** 36 **24.** -20 **25.** Pentagon

26. No, all sides are not segments.

27. Hexagon **28.** No, not closed

29. Quadrilateral or trapezoid

30.
```
 <-+--+--+--+--+--+--+--+--+->
  -3 -2 -1  0  1  2  3  4  5
```

31. $-8, -3, 1, 5$ **32.** $-1, 0, |-2|, |3|$

33. $-4, |1|, |-2|, 3$ **34.** $4s + 4a + 24$; 52

35. $7s + 7a + 7m + 7$; 77 **36.** $3m + 15a$; 39

37. $18a - 6m$; 18 **38.** 40 **39.** $4r - 2n - 5$; 11

40. $8r + 13p - 4n$; 71 **41.** $40r - 20$; 220

42. $-7p + 2n$; -13 **43.** $3rn + 2n + |-p|$; 83

44.–47. See graph.

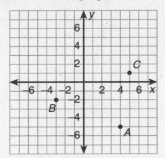

44. Quadrant 4 **45.** Quadrant 3

46. Quadrant 1

47. Answers will vary. Any point in Quadrant 2; $(-x, y)$

48. $a = 14$ **49.** $z = 32$ **50.** $b = -9$

51. $y = 7$ **52.** $c = 24$ **53.** $d = -14$

54. $e = 64$ **55.** $t = -51$ **56.** $f < 11$

57. $g > 19$ **58.** $n \le 16$ **59.** $h \le 7$

60. $i \ge 54$ **61.** 16 ft **62.** 8 in. **63.** 64 sq in.

64. $x(\$0.49 + \$0.35)$ or $\$0.49x + \$0.35x$

65. $2.52 **66.** $1.40

67. $(-2, 1); (-5, -1); (-5, 7); (-2, 7)$

68. Quadrilateral or rectangle **69.** 2

70. Perimeter: 18 units; area: 18 square units

71.–73. Check students' graphs.

74. $3 = n + 15, n = -18$

75. $n - 8 \le 24$; $n \le 32$

76. $9n > 144$; $n > 16$ **77.** $\frac{n}{7} = 8$; $n = 56$

78. $b = 15 + 43 - 49 + 12$

79. $21 **80.** Yes

81.

s	1	2	3	4	5	6	7
l	13	12	11	10	9	8	7

s	8	9	10	11	12	13
l	6	5	4	3	2	1

■ **Short Quiz 4.2**

1. $x = 7$ **2.** $y = -6$ **3.** $z = -8$

4. $a = 39$ **5.** $y = 2$ **6.** $w = 7$

7. Answers will vary. One possible answer: the value of 3 coins and 5¢ = 80¢. $x = 25$

8. $6x + 8x - 2 = 68$ **9.** $x = 5$; 5 tickets.

1. $b = 7$ 2. $c = 16$ 3. $d = 10$
4. $f = -10$ 5. $g = 4$ 6. $h = 1$
7. $j = -2$ 8. See students' sketches.
9. $2w + 2(w + 3) = 4w + 6$
10. Answers will vary but will be ≥ 1.

■ **Mid-Chapter Test, 4-A**

1. $w = 6$ inches, $l = 10$ inches   2. $-\frac{3}{4}$
3. Answers will vary but must be negative and reciprocals of each other.
4. $e = 6$ 5. $y = 3$ 6. $b = 8$
7. $p = -1$ 8. $n = 5$
9. $4x + 7 = 27; x = 5$
10. $22 = 3f - 50; f = 24$
11. $x + (2x - 6) = 495$; 167 plain, 328 with cheese

■ **Mid-Chapter Test, 4-B**

1. $l = 9$ feet, $w = 15$ feet   2. -8
3. Answers will vary but must be positive and reciprocals of each other.
4. $n = 16$ 5. $x = 5$ 6. $r = 12$
7. $y = 4$ 8. $y = -3$
9. $-17 = 10 + 9h; h = -3$
10. $3x - 8 = 25; x = 11$
11. $(35 + x) + x = 365$; old holds 165 barrels, new holds 200 barrels

■ **Short Quiz 4.6**

1. $x = 4$ 2. $t = -1$ 3. $w = -2$
4. $5g + 1 = 4g + 5; g = 4$
5. One possible answer:

6.

Weeks)	Beg. Bal.	1	2	3	4	5	6
Tim	12	15	18	21	24	27	30
Dom	60	55	50	45	40	35	30

7. $12 + 3x = 60 - 5x$ 8. $x = 6$; 6 weeks

■ **Short Quiz 4.8**

1. $x = -2.5$ 2. $f = -0.375 \approx -0.38$
3. $t = 10.94$ 4. $x = 1.80$ 5. 17.6 ft
6. Answers will vary (should be positive)
7. $x = 21$
8. Angle $A = 36°$; Angle $B = 144°$

■ **Chapter Test, 4-A**

1. Multiplying a number by 7
2. Subtracting -5.2 from a number
3. $x = 6$ 4. $y = -16$ 5. $b = 4$
6. $\frac{6}{5}$ 7. $-\frac{1}{4}$ 8. 10   9. $c = -8$
10. $d = 18$ 11. $e = -18$ 12. $f = 6$
13. $5r + 14 + 3r = 2r + 38$ 14. $r = 4$
15. $m = \frac{1}{3}$ 16. $h = 44$ 17. $n = -2.5$
18. $y = 8$ 19. $g = 12$ 20. $x = 21$
21. $A = 56°, B = 34°$ 22. Answers will vary.
23. $v = 2$ 24. Perimeter = 8 units
25. $h = 8.94$ 26. $u = 15.9$ 27. $w = -5.44$
28. 30 minutes from your friend's starting time (20 minutes after your starting time)
29. 24 laps
30.

Minutes	5	10	15	20	25	30
Your laps	0	0	6	12	18	24
Friend's laps	4	8	12	16	20	24

■ **Chapter Test, 4-B**

1. a   2. c   3. c   4. b   5. d   6. d
7. a   8. b   9. b   10. d   11. a   12. d
13. c   14. a   15. d   16. a   17. d   18. b
19. c   20. b   21. d   22. b   23. c   24. d

■ **Chapter Test, 4-C**

1. Adding 7 to a number and multiplying the sum by 3.
2. Dividing a number by 6 and subtracting 5 from the quotient.
3. $4n + 13 = 69; n = 14$
4. $32 + 7n = -3; n = -5$
5. $\frac{1}{3}n - 12 = 5; n = 51$
6. Answers will vary.   7. $x = 31$
8. $A = 62°; B = 57°, C = 126°, D = 115°$

9. -5 **10.** $\frac{1}{32}$ **11.** $s = \frac{3}{5}$ **12.** $x = -32$

13. $n = -\frac{60}{7}$ **14.** $a = 21$ **15.** $y = -32$

16. $\$3.50x + \frac{1}{2}x(\$2.50) = \$28.50; x = 6$

17. $\$13.50$ **18.** $x = 7$

19. When simplified, the equation is the same on both sides. Therefore, it is true for all values of n.

20. Equations may vary. One possible equation:
$$\frac{15x}{2} = 5x + 30; x = 12$$

21. Answers may vary. One possible answer: x represents the number of boxed cookies.

22.

Time (hrs)	$\frac{1}{4}$	$\frac{1}{2}$	$\frac{3}{4}$	1	$1\frac{1}{4}$	$1\frac{1}{2}$	$1\frac{3}{4}$
You	2	4	6	8	10	12	14
Your brother	0	0	3	6	9	12	15

23. After $1\frac{3}{4}$ hours (anytime after $1\frac{1}{2}$ hours when they are tied.)

24. $x = 2.34$ **25.** $e = 0.71$

26. Design costs + cost for first hundred + cost for subsequent hundreds $\leq \$50$

27. You can have 600 notices printed.

28. Solution strategies and explanations will vary.

■ Short Quiz 5.2

1. $\$500$ **2.** $\$2750$ **3.** $\$750$

4. There would be twice as many disks

5. Problems and solutions will vary.

6. 80–89   **7.** 50–59 and 60–69

8. 38   **9.** Answers will vary.

■ Short Quiz 5.4

1. Horizontal axis - time (in weeks); vertical axis, the price of the stock (in dollars)

2. Three different stocks   **3.** RVR

4. Between weeks 1 and 2 and weeks 4 and 5

5. Answers will vary; accept all reasonable responses.

6. Answers will vary; accept all reasonable responses.

7. Graphs will vary but should reflect data.

8. Answers will vary; Possible conclusions: very few kids watch TV for less than 5 hours or more than 20 hours.

■ Mid-Chapter Test, 5-A

1. 1994–1995   **2.** 1993–1994

3. Answers will vary. One possibility: 1993 due to the steady increase in attendance. Accept all reasonable responses.

4. Graphs will vary (line graph or histogram are logical choices).

5. Cola

6. It would still be about 5 less than cola.

■ Mid-Chapter Test, 5-B

1. Popcorn   **2.** Soft drinks

3. Answers will vary.

4. Graphs will vary. (Line graph or histogram are logical choices.)

5. Apples

6. It would still be about 5 less than apples.

■ Short Quiz 5.6

1. $3\frac{1}{2}$ times as many pencils were sold as pens.

2. About 13 more pencils were sold than pens.

3. See students' graphs.

4. Pen salesmen – the graph that students made; pencil salesmen – the misleading graph. The misleading graph makes it appear that many more pencils are sold than pens.

5. See students' line plots.

6. Lions and sharks

■ Short Quiz 5.8

1. 45

2. Cocoa sales rise as the season progresses. (Dates are later.) (Students will hopefully add that this is due to dropping outside temperatures.)

3. See students' sketches.   **4.** About 70.

5. $\frac{7}{16}$ **6.** $\frac{1}{4}$ **7.** $\frac{3}{16}$ **8.** $\frac{1}{8}$ **9.** 1

10. See students' spinners.

11. See students' spinners.

12. See students' spinners.

■ Chapter Test, 5-A

1. Wild turkey - 15 mph; elephant - 25 mph; grizzly bear - 30 mph; zebra - 40 mph; lion - 50 mph; cheetah - 70 mph

2. 45 mph   **3.** About 1787

4. Any events from 1800 forward.

5. Graphs and explanations will vary. (Line plot or bar graph are logical choices.)

6. 9 times as many in 8 and under than in 15–18; really only 8 more in 8 and under than in 15–18.

7. Check students' graphs – scale should reflect numbers 0–18.

8. Check students' line plots.

9. 29   **10.** 7

11. Less – there are far fewer boxes with more than 30 raisins than with 30 or less.

12. Check students' scatter plots – axes should read age (years) and diameter (inches).

13. Positive   **14.** About 9.5 inches   **15.** $\frac{1}{4}$

16. $\frac{2}{3}$ **17.** $\frac{1}{12}$ **18.** $\frac{1}{3}$ **19.** $\frac{2}{3}$

20. Answers will vary. Possible answers: 1 or 3, greater than 4, less than 3, etc.

21. $\frac{1}{2}$

■ Chapter Test, 5-B

1. b   **2.** d   **3.** c   **4.** a   **5.** a

6. a   **7.** a   **8.** b   **9.** c   **10.** b

11. d   **12.** a   **13.** b   **14.** b   **15.** c

16. c   **17.** b   **18.** b   **19.** a   **20.** d

■ Chapter Test, 5-C

1. 10 students

2. Bus: 80; bike: 55; car: 15; walk: 25

3. 25 students   **4.** 40 yrs   **5.** About 1540

6. About 1915   **7.** See students' time lines.

8. Answers will vary – any events from 1800 to the present.

9. Hoover Dam is more than two times as high as Grand Coolee Dam.

10. About 175 feet

11. See students' graphs – scale should be corrected.

12. Answers will vary. One possible answer: Feather and Hoover Dams, since they look significantly higher than the others.

13. See students' scatter plots.   **14.** Positive

15. Answers will vary. Possible answer: 10.5–11.

16. See students' line plots.

17. 8   **18.** $\frac{1}{2}$ **19.** $\frac{1}{4}$ **20.** $\frac{1}{4}$

21. No; some people would move to the next age category.

22. Graphs will vary. Possible choice: line graph since it shows progress through the passage of time.

23. Problems/solutions will vary.

■ Short Quiz 6.2

1. 2, 3, 4, 6   **2.** All

3. 1, 2, 3, 4, 6, 7, 12, 14, 21, 28, 42, 84

4. 1, 2, 3, 4, 6, 8, 12, 16, 24, 48

5. $2^5 \cdot 3$ **6.** $2^3 \cdot 3^3$

7. $-1 \cdot 2 \cdot 2 \cdot 2 \cdot 7 \cdot y \cdot y \cdot z$; $-1 \cdot 2^3 \cdot 7 \cdot y^2 \cdot z$

8. -108

9. Answers will vary; may include 20 rows of 5; 10 rows of 10; 2 rows of 50, and so on.

■ Short Quiz 6.4

1. 24   **2.** $3xy$

3. Answers will vary; possible answer: 15 and 30

4. Answers will vary; may have dimensions of 6×3, 9×2, or 18×1

5. Answers will vary; must match dimensions from Exercise 4

6. No; the area (18) has a factor of 2 and the perimeter (18, 22, or 38) has a factor of 2.

7. 8: 8, 16, 24, 32, 40   **8.** 36: $2^2 \cdot 3^2$
12: 12, 24, 36, 48, 60   54: $3^3 \cdot 2$
LCM: 24   LCM: 108

9. $16xy^2 = 2 \cdot 2 \cdot 2 \cdot 2 \cdot x \cdot y \cdot y$
$20y^4 = 2 \cdot 2 \cdot 5 \cdot y \cdot y \cdot y \cdot y$
LCM: $80xy^4$

■ Mid-Chapter Test, 6-A

1. Yes: 2, 3, 4, 6, 8, 9   No: 5 and 10

2. Yes: 2, 4, 5, 8, 10   No: 3, 6, 9

3. 1, 2, 3, 4, 5, 6, 8, 10, 12, 15, 16, 20, 24, 30, 40, 48, 60, 80, 120, 240

4. $2^5 \cdot 3$ **5.** $2^3 \cdot 3 \cdot 5$ **6.** 5

7. Answers will vary; possible answer: 6 and 12

8. 45   **9.** $\frac{1}{6}$ **20.** $\frac{6}{11}$ **11.** $\frac{x}{3y}$

12. 24 minutes

13. You will have walked 4 laps, your friend will have walked 3 laps.

Mid-Chapter Test, 6-B

1. Yes: 2, 3, 6, 9   No: 4, 5, 8, 10
2. Yes: 2, 4, 5, 10   No: 3, 6, 8, 9
3. 1, 2, 3, 4, 5, 6, 10, 12, 15, 20, 25, 30, 50, 60, 75, 100, 150, 300
4. $2^3 \cdot 3^2$ 5. $3^2 \cdot 7 \cdot 2$
6. Answers will vary; possible answer: 24 and 48
7. 60
8. Answers will vary; possible answer: 3 and 4
9. $\dfrac{1}{7}$ 10. $\dfrac{6}{11}$ 11. $\dfrac{3a}{7b}$ 12. 40 in.
13. 5 8-inch tiles; 4 10-inch tiles

Short Quiz 6.6

1. GCF: 4; $\dfrac{2}{5}$ 2. GCF: 12; $\dfrac{3}{8}$
3. $\dfrac{8}{11}$ 4. $\dfrac{9et}{100}$
5. $\dfrac{1}{4}, \dfrac{3}{16}, \dfrac{5}{8}, \dfrac{1}{2}, \dfrac{21}{32}$; fractions and subsequent orders will vary.
6. $-\dfrac{17}{4}$ 7. Rational; $0.\overline{4}$, repeating
8. $\dfrac{3}{8}$ 9. $5\dfrac{3}{8}$; $\dfrac{43}{8}$; 5.375

Short Quiz 6.8

1. -216 2. $\dfrac{3}{x^9}$ 3. 4   4. 9
5. 3.53×10^5 6. 8.1×10^{-4}
7. 4,160,000   8. 0.00549   9. 9.1
10. Answers vary.

Chapter Test, 6-A

1. Yes: 2, 3, 4, 6, 9   No: 5, 8, 10
2. Yes: 2, 5, 10   No: 3, 4, 6, 8, 9
3. Prime   4. Composite; 1, 2, 4, 23, 46, 92
5. Yes: 12, 16, 17, 18   No: 11, 13, 14, 15, 19, 20
6. $2^2 \cdot 3 \cdot 17$ 7. $5^2 \cdot 7$
8. $2^3 \cdot 3 \cdot a \cdot a \cdot b$ 9. $2^6 \cdot y \cdot y \cdot y$
10. GCF = 5; LCM = 175
11. GCF = $4e$; LCM = $24de^2$
12. 2, 4, 13, 26, or 52   13. $\dfrac{1}{15}$ 14. $\dfrac{m}{5n}$
15. $\dfrac{7}{31}$ 16. Rational; -0.875; terminating

17. Rational; 0.27777. . . ; repeating
18. Irrational; 4.1231056. . . ; non-terminating
19. -64 20. 1   21. $\dfrac{1}{f^4}$ 22. 9   23. 125
24. $24m^6n^2$ 25. $\dfrac{9s^2}{t^5}$ 26. 3400
27. 2,380,000; 2.38×10^6 28. 9.3×10^7
29. To get the next number of dots: add 2, add 3, add 4, etc., to the preceding number of dots; 10 dots, 15 dots, 21 dots.
30. Dimensions: 1×28, 2×14, 4×7; Perimeters: 58, 32, 22

Chapter Test, 6-B

1. c   2. b   3. c   4. a   5. a
6. d   7. b   8. c   9. c   10. d
11. c   12. b   13. c   14. b   15. c
16. a   17. b   18. c   19. d   20. a
21. b   22. a   23. b   24. d   25. d

Chapter Test, 6-C

1. 7   2. 2   3. 2 and 4
4. Possible answer: The number ends in 0 or 5 and the sum of its digits is divisible by 3.
5. 1, 2, 3, 5, 6, 10, 15, 30   Possible dimensions include $2 \times 5 \times 3$; $1 \times 10 \times 3$; $5 \times 6 \times 1$, etc.
6. $2^2 \cdot 3 \cdot 5$ 7. $2^6 \cdot 3$
8. $2 \cdot 2 \cdot 2 \cdot 2 \cdot 2 \cdot a \cdot b \cdot b \cdot b \cdot b \cdot c \cdot c \cdot c \cdot c \cdot c$; $2^5 \cdot a \cdot b^4 \cdot c^5$
9. $-1 \cdot 2 \cdot 3 \cdot 3 \cdot 3 \cdot t \cdot t \cdot u \cdot v \cdot v \cdot v$; $-1 \cdot 2 \cdot 3^3 \cdot t^2 \cdot u \cdot v^3$
10. 5; 120   11. $2xy$; $6x^2y^2$
12. $5abc^2$; $90ab^2c^3$
13. Answers will vary; $\dfrac{9}{16}$ and $\dfrac{17}{32}$. One possible strategy: convert both fractions to common denominators that are multiples of the least common denominator $\left(\dfrac{8}{16} \text{ and } \dfrac{10}{16}; \dfrac{16}{32} \text{ and } \dfrac{20}{32}\right)$, then find fractions between the existing fractions.
14. x^6 15. $\dfrac{s^2}{r^3}$ 16. $-\dfrac{5a}{b^2}$
17. $\dfrac{72g^5h^5}{i^2}$ 18. $-\dfrac{14c^3e^2}{3d^2}$ 19. $\dfrac{4}{9}$
20. 4.9805×10^{-4}; 0.00049805
21. Rational; $0.8\overline{3}$; repeating
22. Rational; 15.0; terminating

23. Irrational; 7.0710678 . . . ; nonrepeating
24. Possible answer: $\frac{1}{2}$ is white, $\frac{1}{4}$ is white, $\frac{1}{8}$ is white. Pattern reflects a fraction whose denominator is a power of 2. Next three terms will be $\frac{1}{16}, \frac{1}{32}, \frac{1}{64}$.
25. 4708; 2.8248×10^5 26. 12620; 7.572×10^5
27. 11960; 7.176×10^5
28. Vapor (4708), brick (11960), hardwood (12620)

■ **Cumulative Test, 1–6**

1. To get the next figure, going clockwise, shade the 2nd triangle after the one shaded in the preceding figure. $6(x - 3)$, 12

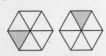

2. To get the next figure, going clockwise, shade the 3rd little square after the one shaded in the preceding figure. $4(2x + 5)$, 60

3. 32768   4. 0.24   5. 8.06   6. 32
7. 17.70   8. 487.97   9. 299
10. Answers will vary; possible answer: $2(x + 3) = 2x + 6$, 16
11. Subtraction, then multiplication by $\frac{1}{3}$ or division by 3
12. -900 13. -14 14. 1   15. 4   16. 10
17. -13 18. $w - 4 = -5; w = -1$
19. $9 = \dfrac{108}{t}; t = 12$
20. $29 + f = -2; f = -31$
21. $14e > -126; e > -9$
22. $-3 \geq 5 + z; z \leq -8$
23. $7 \geq \dfrac{b}{16}; 112 \geq b$ 24. $4b - 4a + 4c$; 68
25. $-5a + 15b - 5c$; 25
26. $|-b| + 4a + 4c$; 32
27. $5 - 2b + c$; 7   28. $-a + 4b - 4c$; -21

29.–32. See graph.

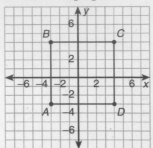

29. Quadrant 3   30. Quadrant 2
31. Quadrant 1   32. Quadrant 4; $(4, -3)$
33. Perimeter: 28 units; area: 49 square units
34. 1   35. $\frac{1}{5}$ 36. -7 37. $t = 17$
38. 5   39. 120   40. $g = 105$
41. $(6y + 6) + (5y + 1) + (4y - 1) + (3y - 6) = 360$
42. $y = 20$ 43. 126°, 101°, 79°, and 54°
44. 3   45. -13 46. -3 47. $-\frac{1}{2}$ 48. 1
49. 4   50. $x = 0.68$ 51. $j = -1.65$
52. Width: 12 in.; perimeter: 40 in.
53. Answers will vary; possible answer: 48 in. by 2 in., 100 in.
54. Positive correlation; the y-coordinates increase as the x-coordinates increase.
55. As the age increases, so does the height.
56. 147 cm   57. About 162 cm
58. Graphs and explanations will vary.   59. Greg
60. Answers will vary. Todd's scores go up and down, so he might get a 95. Greg's score will probably go up to 100, because his score keeps going up.
61. $\frac{5}{12}$ 62. $\frac{1}{6}$ 63. $\frac{1}{6}$ 64. $\frac{1}{4}$ 65. $\frac{7}{12}$
66. GCF: 6; LCM: 36   67. GCF: 60; LCM: 840
68. GCF: $2xy^3$; LCM: $36x^2y^4$
69. GCF: $4a^2b$; LCM: $48a^3b^5$ 70. $\frac{7}{8}$; 0.875
71. $\frac{1}{3}$; 0.333   72. $\frac{32}{33}$; 0.970   73. 7.348
74. Answers will vary; possible answer: $14ab, 7a^3b$
75. Answers will vary; possible answer: $10c, 4c^2$
76. $\dfrac{1}{27}$ 77. 1   78. $\dfrac{1}{x^4}$ 79. 216   80. y
81. $42b^4$ 82. $\dfrac{5}{3e}$ 83. 1.9655×10^6
84. 163,791.67; 1.6379167×10^5 85. 5.64×10^{-5}

Short Quiz 7.2

1. $-\dfrac{3}{7}$ 2. $2\dfrac{1}{3}$ 3. $\dfrac{11y}{12}$ 4. $-\dfrac{3}{4r}$

5. $-\dfrac{9}{11}$ 6. 1   7. $1\dfrac{1}{12}$ 8. $\dfrac{1}{24y}$

9. $\dfrac{5+6t}{st}$ 10. $\dfrac{54+4c}{9c}$ 11. Answers will vary.

Short Quiz 7.4

1. $0.471x - 0.438x$; $0.03x$

2. $0.375 + 0.571 + 0.563 - 1$; 0.51

3. Answers will vary; ranging from 2.885 to 2.894.

4. $-\dfrac{9}{25}$ 5. $-\dfrac{529}{48}$ 6. $\dfrac{8w}{5}$

7. $-\dfrac{3}{2}$ 8. 8.531 square units

Mid-Chapter Test, 7-A

1. $\dfrac{7}{16}$ 2. $\dfrac{82}{75}$ or $1\dfrac{7}{75}$ 3. $-\dfrac{5}{32}$

4. $-2a$ 5. $d = \dfrac{1}{24}$ 6. $x = \dfrac{11}{4}$

7. Answers will vary.   8. 0.47

9. $\dfrac{1}{3}$; $\dfrac{1}{5}$; $\dfrac{1}{15}$

10. $48.95 at $\dfrac{1}{3}$ off; $\sim$2.31

Mid-Chapter Test, 7-B

1. $\dfrac{1}{12}$ 2. $\dfrac{71}{70}$ or $1\dfrac{1}{70}$ 3. $-\dfrac{27}{175}$

4. $-\dfrac{4n}{9}$ 5. $d = -\dfrac{13}{45}$ 6. $k = \dfrac{1}{6}$

7. Answers will vary.   8. 8.41

9. $\dfrac{3}{8}$; $\dfrac{2}{3}$; $\dfrac{1}{4}$

10. 2 pairs each at $\dfrac{1}{5}$ off; $5.50

Short Quiz 7.6

1. $\dfrac{3}{2}$ 2. $-\dfrac{15}{56}$ 3. $-\dfrac{x}{25}$ 4. 8

5. 18 s'mores   6. 36%   7. 46%

8. 16%   9. Answers will vary.

10. Answers will vary.

Short Quiz 7.8

1. 0.435   2. 0.275   3. 84.3%   4. 167%

5. $\dfrac{7}{25}$ 6. $\dfrac{11}{10}$ 7. 0.36; 12.96   8. 1.25; 200

9. Answers will vary. Be sure that the second rectangle has dimensions 50% smaller than the original rectangle.

10. Yes, the perimeter of the smaller rectangle is 50% of the perimeter of the larger rectangle; $2(0.5l) + 2(0.5w) = 0.5(2l + 2w)$

Chapter Test, 7-A

1. $\dfrac{17}{22}$ 2. $\dfrac{3}{16}$ 3. $\dfrac{2}{5}t$ 4. $-\dfrac{4+9a}{ab}$

5. $\dfrac{19}{18}$ or $1\dfrac{1}{18}$ 6. 12   7. $\dfrac{15m}{2}$ 8. $-\dfrac{1}{3}$

9. $-\dfrac{1}{2} = z$ 10. $t = \dfrac{11}{45}$ 11. $b = \dfrac{41}{28}$ or $1\dfrac{13}{28}$

12. $u = \dfrac{1}{40}$ 13. $q = \dfrac{15}{7}$ or $2\dfrac{1}{7}$

14. $t = -3$ 15. 11

16. Perimeter: $\dfrac{114}{10}$ or $11\dfrac{2}{5}$ cm

    Area: $\dfrac{351}{50}$ or $7\dfrac{1}{50}$ sq cm

17. 1.10   18. $-0.42y$ 19. 0.29   20. -1.14

21. Answers will vary.   22. 4.2%

23. $\dfrac{146}{100}$; $\dfrac{73}{50}$ or $1\dfrac{23}{50}$ 24. 23   25. 165

26. Total area: 10 square units
    Shaded: 70%
    Striped: 10
    Dotted: 20%

27. 9 hours

28. Answers will vary.   29. $10.87

30. About 25.5%   31. About 27.7%

32. About 16.4%   33. Answers will vary.

Chapter Test, 7-B

1. d   2. c   3. a   4. b   5. a   6. b

7. b   8. a   9. d   10. c   11. b   12. d

13. c   14. a   15. b   16. c   17. d   18. c

19. b   20. d   21. a   22. b   23. d   24. b

25. d   26. c

Chapter Test, 7-C

1. $\dfrac{23}{40}$ 2. $\dfrac{49}{207}$ 3. $\dfrac{3(2-a)}{ab}$ or $\dfrac{6-3a}{ab}$

4. $\dfrac{-5fg-48}{8fg}$ 5. $\dfrac{63}{20}$ or $3\dfrac{3}{20}$ 6. $13\dfrac{7}{76}$

7. 1   8. $-2x$ 9. $x = \dfrac{5}{16}$

10. $j = \dfrac{28}{525} = \dfrac{4}{75}$ 11. $y = -\dfrac{3}{2}$ or $-1\dfrac{1}{2}$

12. $w = -\dfrac{29}{8}$ or $-3\dfrac{5}{8}$

13. Perimeter: $28\dfrac{7}{10}$ inches

    Area: $22\dfrac{1}{20}$ square inches

14. $15t + \frac{4}{5} = -\frac{9}{10}$; $t = -\frac{17}{150}$

15. $\frac{v}{30} = \frac{2}{3}$; $v = 20$ 16. -0.44 17. 0.25

18. 39.47% 19. 242.86% 20. 58.9%

21. 7.34% 22. 0.027 23. Answers will vary.

24. 30.45 25. 68.8%

26. Total area: $1\frac{11}{16}$ square units
    White: $11.\overline{1}\%$
    Shaded: $33.\overline{3}\%$
    Starred: $\approx 25.9\%$
    Striped: $\approx 29.6\%$

27. Answers will vary. Percents, fractions and decimals should reflect the unshaded portion of the figure and should be equivalent.

28. $\approx 55.3\%$ are NOT orange

29. $\approx 47.4\%$ grow less than four feet

30. 17 sunflowers   31. About 39 students

32. About 98 students

■ **Short Quiz 8.2**

1. Rate – two different units of measure

2. $\frac{660}{5280}$; $\frac{1}{8}$ 3. $\frac{1000}{500}$; $\frac{2}{1}$

4. Answers will vary but should be less than $1.19.

5. $x = 32$ 6. $x = 6$ 7. $\frac{t}{5} = \frac{60}{25}$; $t = 12$

8. 14.12 9. $b = 12, c = 13$

■ **Short Quiz 8.4**

1. 26 tickets   2. 10,530 seats   3. $\approx \$10$

4. 13% 5. 2.88 6. 90 7. 500 8. 8

9. Answers will vary. The percent should be exactly twice the number of heads tossed.

■ **Mid-Chapter Test, 8-A**

1. Rate: $\frac{18 \text{ wheels}}{1 \text{ truck}}$ 2. Rate: $\frac{1 \text{ class}}{29 \text{ students}}$

3. Ratio: $\frac{12}{13}$ 4. $\frac{o}{r} = \frac{p}{s}$; $\frac{p}{s} = \frac{q}{t}$; $\frac{o}{r} = \frac{q}{t}$

5. Yes   6. No   7. 334   8. 72

9. 14% 10. $x = 2$ 11. $z = 5$

12. Answers will vary. Percent should not be greater than 100% and the number of students should not exceed 700.

■ **Mid-Chapter Test, 8-B**

1. Rate: $\frac{9 \text{ appointments}}{1 \text{ day}}$ 2. Ratio: $\frac{1}{4}$

3. Rate: $\frac{1 \text{ tank}}{12 \text{ gallons of gas}}$

4. $\frac{a}{b} = \frac{c}{d}$; $\frac{a}{b} = \frac{e}{f}$; $\frac{c}{d} = \frac{e}{f}$

5. No   6. Yes   7. 225   8. 75

9. 25% 10. 48   11. $7.\overline{27}$

12. Answers will vary. Number of shots should not exceed 24 and the percent should not be greater than 100%.

■ **Short Quiz 8.6**

1. 500   2. 125   3. 100   4. 855

5. Increase; 25% 6. Decrease; 24%

7. Answers will vary.   8. 1102

■ **Short Quiz 8.8**

1. 15   2. 12

3.

4. 35 ways   5. 32

6. Answers will vary. One possible answer: even-H

7. $\frac{2}{36}$ or $\frac{1}{18}$ 8. $\frac{5}{36}$

■ **Chapter Test, 8-A**

1. Ratio: $\frac{20}{1}$ 2. Rate: $\frac{12 \text{ campers}}{1 \text{ bunk}}$

3. $\frac{46}{50}$, ratio, $\frac{23}{25}$ 4. $\frac{2 \text{ buses}}{112 \text{ students}}$, rate, $\frac{1 \text{ bus}}{56 \text{ students}}$

5. $c = \frac{35}{12}$ 6. $d = 45$ 7. $y = 30$

8. $v = \frac{136}{3}$ or $45\frac{1}{3}$ 9. $16.\overline{6}\%$ 10. ≈ 72.8

11. 67.5 12. 27 13. $\frac{3.2}{1}$ or $\frac{16}{5}$ 14. $v = 4$

15. $w = 5$ 16. $\frac{16}{5}$ 17. Increase; $\approx 42.9\%$

18. Decrease; 62% 19. Answers will vary.

20. White van, blue van, red van, black van, gold van, white truck, blue truck, red truck, black truck, gold truck, white station wagon, blue station wagon, red station wagon, black station wagon, gold station wagon.

21. $3 \cdot 5 = 15$ 22. $\frac{5}{12}$ 23. $\frac{1}{4}$ 24. $\frac{1}{6}$ 25. $\frac{1}{3}$

26. See students' sketches.

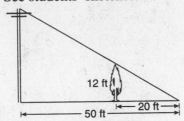

12 ft

20 ft

50 ft

27. $\dfrac{x}{12} = \dfrac{50}{20}$ **28.** 30 ft **29.** $35.97 **30.** $9.38

■ **Chapter Test, 8-B**

1. b **2.** d **3.** b **4.** d **5.** c **6.** a
7. d **8.** a **9.** d **10.** b **11.** a **12.** b
13. c **14.** c **15.** d **16.** a **17.** a **18.** b
19. c **20.** d **21.** b **22.** c **23.** d **24.** c
25. c **26.** a **27.** d

■ **Chapter Test, 8-C**

1. Rate: $\dfrac{11 \text{ players}}{1 \text{ game}}$ **2.** Ratio: $\dfrac{43}{34}$

3. Answers will vary. **4.** Answers will vary.

5. $y = \frac{7}{3}$ **6.** $z = \frac{48}{7}$ **7.** $a = 7$

8. Answers will vary. **9.** 28%

10. 200 **11.** 42.25 **12.** $\frac{5}{3}$

13. $\frac{40}{3}$ or $v = 13\frac{1}{3}$ **14.** $\frac{50}{3}$ or $w = 16\frac{2}{3}$

15. Increase, 6% **16.** Increase, about 3.27%

17. Explanations and estimates will vary; $7.50.

18. 117 **19.** $\dfrac{750}{x} = \dfrac{37.5}{31}$ **20.** 620 feet

21. $55.\overline{5}\%$ **22.** $33.\overline{3}\%$ **23.** $\frac{1}{9}$

24. 24 **25.** 21 **26.** $\frac{1}{2}$ **27.** $\frac{1}{13}$

28. No; The probability of drawing a heart is $\frac{13}{52}$ or $\frac{1}{4}$ and the probability of drawing a face card is $\frac{12}{52}$ or $\frac{3}{13}$. $\frac{13}{52} > \frac{12}{52}$

29. 8% raise; 8% of $4.75 is $0.38 which is greater than the $0.30 raise.

■ **Short Quiz 9.2**

1. $+9, -9$ **2.** $+\frac{3}{2}, -\frac{3}{2}$ **3.** $y = +7, -7$

4. $x = +8, -8$ **5.** $t^2 - 1 = 24; t = +5, -5$

6. Estimates may vary. Accept answers close to 5.5 units.

7. Answers will vary.

8. $3 + \sqrt{3}$; no **9.** $\frac{3}{5}$; yes

■ **Short Quiz 9.4**

1. $c = 8.944$ **2.** $b = 12$

3. $a = 17.861$ **4.** $t \approx 15.81$

5. Perimeter = 42 units; area = 108 square units

✓ **6.** 36 feet

7. Answers will vary, but numbers should be multiples of 1.5, 2, 2.5

■ **Mid-Chapter Test, 9-A**

1. $+13, -13$ **2.** $+0.2, -0.2$ **3.** 6.504 units

4. Irrational **5.** Rational **6.** $x = 13.675$

7. $y = 26$ **8.** 20 feet away

■ **Mid-Chapter Test, 9-B**

1. $+16, -16$ **2.** $+0.3, -0.3$

3. 6.892 units **4.** Irrational

5. Rational **6.** $m = 8$

7. $n = 13.454$ **8.** 1.3 m or 130 cm

■ **Short Quiz 9.6**

1.

2.

3. $x \le 2$ **4.** $x > -5$

5. $w + 2 < 14; w < 12$

6. $y > -2$

7. $n \le 4$

8. $m > \frac{1}{4}$

9. $x \ge \frac{8}{9}$

10. At least 94 yards of ribbon

■ **Short Quiz 9.8**

1. $y > 1$ **2.** $m \ge 9$ **3.** $g > \frac{11}{2}$

4. $n + n + 1 \le -21; n \le -10$

5. $n + n + 1 + n + 2 > 61; n > 19\frac{1}{3}$

6. Yes; the sum of each pair of sides is greater than the measure of the third side.

7. No, $0.325 + 0.525$ is not greater than 0.875.

8. Answers will vary. $\frac{2}{10} <$ third side $< \frac{8}{10}$

■ **Chapter Test, 9-A**

1. $+25, -25$ **2.** $+1.4, -1.4$

3. $+\frac{7}{9}, -\frac{7}{9}$ **4.** $d = +9, -9$

5. $z = +\sqrt{6}, -\sqrt{6} \approx +2.449, -2.449$

6. $f = +11, -11$

7. C   **8.** A   **9.** B   **10.** D

11. Answers will vary. Check that the number is plotted correctly.

12. 3.513 **13.** 8 feet, ≈ 11.662 feet

14. $b = 32$ **15.** $g = 15$ **16.** $w = 5.657$

17.
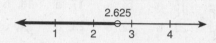

18.

19.

20.

21. $k < -8$ **22.** $y > -1$ **23.** $t \leq \frac{2}{3}$

24. $g \leq 5\frac{3}{5}$ or $\frac{28}{5}$ **25.** $w > \frac{6}{7}$

26. $5 < x < 21$ **27.** $26 <$ perimeter < 42

28. See students' sketches.

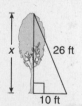

26 ft

10 ft

29. 24 feet

■ **Chapter Test, 9-B**

1. b   **2.** a   **3.** d   **4.** b   **5.** a

6. a   **7.** d   **8.** a   **9.** c   **10.** d

11. a   **12.** b   **13.** c   **14.** b   **15.** a

16. d   **17.** d   **18.** a   **19.** c   **20.** c

21. d   **22.** c   **23.** b   **24.** c   **25.** a

■ **Chapter Test, 9-C**

1. $+23, -23$ **2.** $+0.6, -0.6$

3. $+\frac{11}{15}, -\frac{11}{15}$ **4.** $x = +9, -9$

5. $x = +12, -12$ **6.** $x = +2.3, -2.3$

7. D   **8.** A   **9.** B

10. Answers will vary.   **11.** Answers will vary.

12. 6.509 **13.** $12.062, 3.062$

14. $s = 18$ **15.** $y = 9.434$ **16.** $f = 3.873$

17. Answers will vary. Students should draw and label a right triangle with one side labeled 7 units.

18. $x < 2.625$

2.625

19. $t \geq -84$

20. $b < 0.5$

21. $3s + 7 < 10; s < 1$

22. $14.6 \leq -3q; q \leq -4.8666$

23. $-5(x - 7) < -15; x > 10$

24. 4 video games   **25.** 7 pinball games

26. 25.98 feet   **27.** 16.58 feet

28. See students' sketches.
One possible sketch:

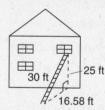

30 ft   25 ft

16.58 ft

29. Yes; $(6.5)^2 + 3^2 > 7^2$

■ **Cumulative Test 7–9**

1. 1   **2.** $\frac{1}{5}$ **3.** $\frac{2y}{3}$ **4.** $\frac{7}{30}$ **5.** $\frac{31}{20}$

6. $\frac{1}{24}w$ **7.** 1.481 **8.** 0.409 **9.** 0.234

10. $6.089g$ **11.** Perimeter: $34\frac{6}{7}$ ft; area: $75\frac{46}{49}$ sq ft

12. Perimeter: $22\frac{19}{20}$ cm; area: $29\frac{9}{20}$ sq cm

13. $\frac{9}{22}$ **14.** $\frac{5}{16}$ **15.** $9r$ **16.** $\frac{3t}{2}$

17. 0.25; 25%   **18.** 0.48; 48%   **19.** $0.7\overline{6}$; $76.\overline{6}\%$

20. 0.71; 71%   **21.** $\frac{11}{25}$; 0.44; 44%   **22.** $569.63

23. $99.75   **24.** 28   **25.** 81   **26.** 26.25

27. 91.4%   **28.** Rate; $\dfrac{6 \text{ swimmers}}{1 \text{ heat}}$

29. Ratio; $\dfrac{4.3}{1}$ **30.** Rate; $\dfrac{3 \text{ months}}{13 \text{ weeks}}$

31. Ratio; $\frac{7}{8}$ **32.** 15,731,400

33. About 12.35   **34.** About 10.616   **35.** 32

36. 7   **37.** $15\frac{3}{4}$ **38.** ≈ 2.14 **39.** 57.75 cm

40. 20%   **41.** 9.72   **42.** 56   **43.** 93.75%

44. 2.7   **45.** 20   **46.** About $111 ($110.94)

47. 6; One possible explanation: 12% of 100 is 12, 50 is half of 100; 6 is half of 12.

48. 250   **49.** 10   **50.** 100   **51.** 30   **52.** 35

53. 15   **54.** 20   **55.** Increase, $\approx 116\%$

56. Decrease, $\approx 20\%$ **57.** Increase, $\approx 23.1\%$

58. $\frac{1}{156}$ **59.** $\frac{1}{78}$ **60.** $\frac{1}{4}$ **61.** $\frac{1}{78}$

62. -1.22

63. 0.94   **64.** -0.92 **65.** 1.73

66. Answers will vary, but the value of the point should match its placement on the number line (and be between 0 and 0.5).

67. $p \approx 20.59$ **68.** $r \approx 6.2$ **69.** $s \approx 24.45$

70. $b < 16$

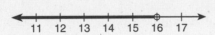

71. $i \geq 16.\overline{3}$

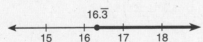

72. $f < 1.8$

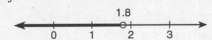

73. $r \leq \frac{9}{2}$

74. $k \geq \frac{4}{9}$

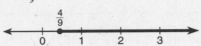

75. No; $7 + 5$ is not greater than 12.8.

76. Yes; $1 + \sqrt{3}$ is greater than $\sqrt{3}$.

77. No, $8 + \sqrt{90}$ is not greater than 18.

78. No, $3 + 4$ is not greater than 7.

79. $r^2 + 14 = 135$, $r = +11, -11$

80. $4y^2 = 64$; $y = 4, -4$

81. $-\dfrac{t}{7} \geq -5$; $t \leq 35$

82. Two possibilities: $c \approx 12.21$, $b \approx 7.14$

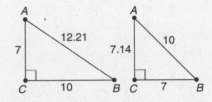

83. $2(3x + 4) + 2(8x - 2) = 48$; $x = 2$

84. Length: 14 inches, width: 10 inches

85. ≈ 17.2 inches   **86.** $3.00

87. 10% off of $25.99 = $2.60,
Sale price = $23.39
15% off of $32.99 = $4.95,
Sale price = $28.04
$25.99 item is less expensive.

88. $51.93

■ Short Quiz 10.2

1. Possible answers: $\overleftrightarrow{US}$, $\overleftrightarrow{UY}$, $\overleftrightarrow{UR}$, $\overleftrightarrow{VY}$, $\overleftrightarrow{VR}$, $\overleftrightarrow{SV}$, $\overleftrightarrow{SY}$, $\overleftrightarrow{SR}$, $\overleftrightarrow{YR}$

2. $\overline{WU}$, $\overline{UT}$, and $\overline{WT}$

3. $\overrightarrow{UV}$, $\overrightarrow{UT}$, $\overrightarrow{UW}$, $\overrightarrow{US}$ or $\overrightarrow{YR}$, $\overrightarrow{YX}$, $\overrightarrow{YZ}$, $\overrightarrow{YU}$

4. $\overleftrightarrow{ZX}$ and $\overleftrightarrow{WT}$

5. $\overleftrightarrow{WT}$ and $\overleftrightarrow{RS}$; $\overleftrightarrow{ZX}$ and $\overleftrightarrow{RS}$

6. $\angle RYZ$, XYV, VUW, and TUS

7. $\angle RYX$, ZYV, VUT, WUS

8. Possible answers: $\angle ZYX$, WUT, RYV, VUS

9. Sketches will vary. Angles should be less than 90°.

10. Sketches will vary. Angles should be greater than 90°.

■ Short Quiz 10.4

1. Lines a and b 2. $\angle s$ 2, 4, 5, and 7
3. Answers should include 2 of the following pairs: $\angle s$ 4 and 2, 1 and 3, 5 and 7, 6 and 8.
4. $\angle s$ 2 and 5, 1 and 6, 3 and 8, 4 and 7
5. $\angle 1$ is congruent to $\angle 4$ (vertical angles); $\angle 4$ is congruent to $\angle 8$ (corresponding angles); $\angle 8$ is congruent to $\angle 2$ (corresponding $\angle s$)
6. Answers will vary. Check students' sketches.
7. Answers will vary. Check students' sketches.
8. HAM

■ Mid-Chapter Test, 10-A

1. F
2. Any two of the following: $\overleftrightarrow{AH}$, $\overleftrightarrow{CF}$, or $\overleftrightarrow{BE}$
3. F 4. $\angle s LIJ, HIL$ 5. $\angle s HIK, MIJ$
6. $\angle s KIJ, LIK, LIM, MIH, MIK$
7. $\angle HIJ$
8. Check students' sketches. All sides should have different lengths and one angle should be greater than 90°.
9. Check students' sketches. Two sides should have the same length and all angles should be less than 90°.
10. Check students' sketches. All sides should have the same length and all angles the same measure (60°).
11. 6 lines of symmetry, rotational symmetry at 60°, 120°, and 180° in either direction
12. r and s are parallel.
13. $\angle s 1$ and 4, 2 and 3, 5 and 8, 6 and 7
14. $\angle s 1$, 4, 5, and 8

■ Mid-Chapter Test, 10-B

1. O
2. Any two of the following: $\overleftrightarrow{QP}$, $\overleftrightarrow{LM}$, or $\overleftrightarrow{NO}$
3. N 4. $\angle s XUZ$ and XUV
5. $\angle s ZUW, WUY, VUY$
6. $\angle s ZUY, YUX, XUW, WUV$ 7. $\angle VUZ$
8. Check students' sketches. All sides should have different lengths and all angles should be less than 90°.

9. Check students' sketches. Two sides should have the same length and one angle should measure 90°.
10. Check students' sketches. All sides should have the same length and all angles should have the same measure (60°).
11. 6 lines of symmetry, rotational symmetry at 60°, 120°, and 180° in either direction
12. a and b
13. $\angle 1$ and 5, 3 and 7, 2 and 6, 4 and 8
14. $\angle 2, 3, 6,$ and 7

■ Short Quiz 10.6

1. Check students' sketches. Two sides should have the same length and one angle should be greater than 90°.
2. Check students' sketches. All sides should have different lengths and one angle should measure 90°.
3. Check students' sketches. All sides should have the same length and all angles should have the same measure (60°).
4. Check students' triangles; right scalene
5. scalene right
6. Parallelogram; check students' sketches.
7. Trapezoid; check students' sketches.
8. Rhombus; check students' sketches.
9. $x = 4.5$, $y = 10$; perimeter: 29 units

■ Short Quiz 10.8

1. Equiangular, equilateral, regular; see students' sketches.
2. Equilateral; see students' sketches.
3. See students' sketches, square
4. 360°   5. 360°
6. No, the sides are different lengths (or the angles have different measures).
7. $m\angle 1 = 90°$, $m\angle 2 = 90°$, $m\angle 3 = 70°$, $m\angle 4 = 70°$, $m\angle 5 = 110°$

■ Chapter Test, 10-A

1. $\overleftrightarrow{MR}$, $\overleftrightarrow{TR}$
2. $\overline{TY}$, $\overline{TR}$, $\overline{TM}$, $\overline{TE}$, and $\overline{TX}$
3. $\overrightarrow{GO}$ 4. $\angle s OET, YEO, TEO$
5. See students' drawings, acute
6. See students' drawings, obtuse

7. Lines t and u

8. ∡1 and 6, 2 and 5, 3 and 8, 4 and 7, 9 and 14, 10 and 13, 11 and 16, 12 and 15

9. ∡1, 6, 3, and 8   10. ∡13, 10, 15, and 12

11. Possible answers: 1 and 3, 2 and 4, 5 and 7, 6 and 8, 9 and 11, 10 and 12, 13 and 15, 14 and 16

12. Possible answers: 1 and 9, 5 and 13, 2 and 10, 6 and 14, 3 and 11, 7 and 15, 4 and 12, 8 and 16

13. Acute isosceles   14. Right scalene

15. See students' sketches, should be an isosceles trapezoid; trapezoid.

16. Equilateral, equiangular and regular octagon; check students' sketches.

17. 8 lines of symmetry, rotational symmetry at 45°, 90°, 135°, and 180° in either direction

18. 42°   19. 48°   20. $\overline{XY}$

21. See students' drawings; right scalene

22. Drawings and explanations will vary. See students' sketches.

23. See students' drawings; interior angles measure 120°, exterior angles measure 60°

24. $\overline{DF}, \overline{FF}, \overline{DF}$ 25. 69°, 69°, (42°)

26. 20°, 134°, (26°)   27. 45°, 45°, (90°)

■ **Chapter Test, 10-B**

1. c   2. d   3. a   4. b   5. a

6. b   7. c   8. a   9. d   10. b

11. b   12. c   13. c   14. b   15. a or B

16. d   17. b   18. b   19. a   20. d

21. d   22. c   23. a   24. a   25. d

■ **Chapter Test, 10-C**

1. $\overleftrightarrow{LS}, \overleftrightarrow{LI}, \overleftrightarrow{SI}, \overleftrightarrow{SN}$, and $\overleftrightarrow{IN}$

2. $\overrightarrow{IN}, \overrightarrow{IW}, \overrightarrow{ID}, \overrightarrow{IS}$ 3. $\overleftrightarrow{WD}$ and $\overleftrightarrow{OA}$

4. $\overleftrightarrow{WD}$ and $\overleftrightarrow{OA}$; they are parallel.

5. Possible answers: ∡ASL and ISO; ∡WIN and SID

6. Possible answers: ∡NID and ISA; ∡WIS and OSL

7. Possible answers: ∡OSA, WID, SIN, LSI

8. Check students' sketches. Angle should be < 90°.

9. Check students' sketches. Angle should be > 90°.

10. Check students' sketches. Angle should measure 90°.

11. Possible answer: a right angle has only one possible measure; acute and obtuse angles have many possible measures.

12. Check students' sketches; two sides should have the same length and all angles should be < 90°.

13. Check students' sketches; one angle should measure 90° and all sides should be different lengths.

14. $x = 7m, y = 12m$

15. Check students' sketches.

16.   17. ∠L, ∠K, ∠J

18. Possible answers: ∡1 and 6, 3 and 8, 9 and 14, 11 and 16, 2 and 5, 4 and 7, 10 and 13, 12 and 15

19. ∠1   20. ∠14

21. ∠13 is larger than ∠14; ∠14 is less than 90°, ∠13 must be greater than 90° since they combine to make a straight (180°) angle.

22. ∡2, 4, and 5   23. 102°   24. 65°

25. 69°   26. 124°   27. 360°

28. $m\angle 1 = 33°, m\angle 2 = 57°, \overline{AL}, \overline{AG}, \overline{GL}$

29. $m\angle 1 = 65°, m\angle 2 = 60°; \overline{GI}, \overline{GH}, \overline{HI}$

■ **Short Quiz 11.2**

1. 26.25

2. 52.5; multiply the area of A by 2 or multiply base times height.

3. See students' sketches. Perimeter and area will vary.

4. About 1000 square meters   5. ∠L 6. $\overline{LI}$

7. $\overline{TP}$ 8. ∠O 9. ∠D 10. $\overline{ID}$

11.

■ Short Quiz 11.4

1.   **2.**

3.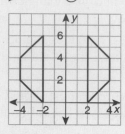

4. Horizontal:
0, 3, 8, possibly 1
Vertical:
0, 8, possibly 1

5. 180° **6.** 135° clockwise

7.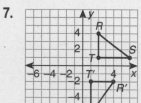

8. 3 units

9. $\angle T'$

10. 6 square units

■ Mid-Chapter Test, 11-A

1. Answers will vary but may be a parallelogram, rectangle, or square.

2. 42 square units **3.** 34 units

4. $\angle P$ **5.** $\overline{AL}$ **6.** $\angle E$ **7.** $\overline{RE}$

8. x-axis (horizontal line) **9.** 90° clockwise

10. 4 units to the left and 2 units down

■ Mid-Chapter Test, 11-B

1. A triangle **2.** 42 square units

3. 34 units **4.** $\angle E$ **5.** $\overline{AN}$

6. $\angle T$ **7.** $\overline{RD}$ **8.** y-axis (vertical line)

9. 60° clockwise

10. 3 units to the right and 4 units down

■ Short Quiz 11.6

1. 4 units to the right; $(x + 4, y)$

2. 1 unit to the right and 3 units down; $(x + 1, y - 3)$

3.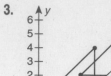

4. See students' sketches.

5. $\dfrac{TR}{ZO} = \dfrac{RA}{OI} = \dfrac{AP}{ID} = \dfrac{PT}{DZ}$

6. 3 to 1 **7.** 18 **8.** $\angle P$

■ Short Quiz 11.8

1. 814 m by 1332 m

2. Answers will vary. One possible answer: 1 cm = 4 ft; model would be 5 cm by 7 cm

3. $\dfrac{7}{\sqrt{74}} \approx 0.814$ **4.** $\dfrac{5}{\sqrt{74}} \approx 0.581$

5. $\dfrac{7}{5} = 1.4$

6. 60°, 20;

$$\sin 30° = \dfrac{1}{2} = 0.5$$

$$\cos 30° = \dfrac{\sqrt{3}}{2} \approx 0.8660$$

$$\tan 30° = \dfrac{1}{\sqrt{3}} \approx 0.5774$$

$$\sin 60° = \dfrac{\sqrt{3}}{2} \approx 0.8660$$

$$\cos 60° = \dfrac{1}{2} = 0.5$$

$$\tan 60° = \dfrac{\sqrt{3}}{1} \approx 1.732$$

7. 54°, 2.8;

$$\sin 36° = \dfrac{2.8}{4.9} \approx 0.5714$$

$$\cos 36° = \dfrac{4}{4.9} \approx 0.8163$$

$$\tan 36° = \dfrac{2.8}{4} = 0.7$$

$$\sin 54° = \dfrac{4}{4.9} \approx 0.8163$$

$$\cos 54° = \dfrac{2.8}{4.9} \approx 0.5714$$

$$\tan 54° = \dfrac{4}{2.8} \approx 1.4286$$

Chapter Test, 11-A

1. Perimeter = $26 + \sqrt{52} \approx 33.21$;
   area = 60 square units

2. Perimeter = 52 units; area = 96 square units

3. $\overline{TI}$ 4. $\overline{BX}$ 5. $\overline{NI}$

6. $\angle T$ 7. $\angle I$ 8. $\angle X$

9. $(3, -2), (5, -1), (5, -4), (3, -6)$

10. $(2, -3), (6, -3), (4, -5), (1, -5)$

11. $(-5, -4), (-3, -5), (-3, -2), (-5, 0)$

12. $(-2, 2), (-4, 1), (-4, 4), (-2, 6)$

13. 60° counterclockwise rotation

14. Reflected across the line

15. translation of 3 units to the right and 2 units down (or 3 units to the left and 2 units up)

16. $NC = 4; LA = 15$ 17. $\dfrac{NY}{LA} = \dfrac{YC}{AX} = \dfrac{NC}{LX}$

18. $\dfrac{5}{12.5}$ or $\dfrac{1}{2.5}$ 19. $\dfrac{15}{37.5}$, yes

20. Check students' sketches. Answers will vary.

21. 24.75 cm   22. 7.5

23. Frame = 334.125 square centimeters;
    drawing = 5.94 square centimeters

24. 56.25; It is the square of the scale factor.

25. 0.92   26. 0.38   27. 2.4   28. 0.38

29. 0.92   30. 0.42   31. $x = 20$

Chapter Test, 11-B

1. a   2. d   3. c   4. c   5. d   6. c

7. d   8. a   9. b   10. a   11. a   12. b

13. a   14. b   15. b   16. a   17. c   18. c

19. b   20. a   21. b   22. a   23. d   24. a

25. c   26. b

Chapter Test, 11-C

1. Perimeter = 42.94 units;
   area = 105 square units

2. Perimeter = 49.6 units;
   area = 148 square units

3. $\overline{CA} \cong \overline{VA}; \overline{AR} \cong \overline{AN}; \overline{RC} \cong \overline{NV}$

4. $\angle C \cong \angle V; \angle A \cong \angle A; \angle R \cong \angle N$

5. $(-1, -2), (-1, -3), (-2, -4), (-4, -2)$

6. $(-1, 2), (-1, 3), (-2, 4), (-4, 2)$

7. $(1, -2), (1, -3), (2, -4), (4, -2)$

8. Answers will vary, but students should note that when the figure is reflected about the x-axis, the x-coordinates are the same and the y-coordinates of the image are the opposite of the y-coordinates of the original figure. The reverse is true when the figure is reflected about the y-axis.

9. $(6, -8), (6, -9), (5, -10), (3, -8)$

10. $(1, 2), (1, 3), (2, 4), (4, 2)$

11. Congruent; 180° clockwise or counterclockwise

12. Congruent; reflected about the line

13. Congruent; translated 4 units to the right and 2 units down

14. $\dfrac{AB}{EF} = \dfrac{BC}{FG} = \dfrac{CD}{GH} = \dfrac{DA}{HE}$

15. $\dfrac{5}{12.5}$ or 0.4   16. 10

17. 10   18. $\dfrac{2}{5} = 0.4$ 19. $\dfrac{4}{25} = 0.16$

20. The sides and the perimeter have a scale factor of 0.4. The area has a scale factor of $(0.4)^2$ or 0.16, since it is measured in square units.

21. 5.2   22. 0.92   23. 0.38   24. 2.4

25. 0.38   26. 0.92   27. 0.42

28. Answers will vary. One possible answer: $\sin T = \cos P$ because it is the quotient of the same two numbers. The opposite side of T (for $\sin = $ opp/hyp) is the adjacent side for P (for $\cos = $ adj/hyp).

29. $x = 7.3; y = 3.4$ 30. $x = 9.6; y = 3.3$

31. 15 feet   32. About 21.2 feet

Short Quiz 12.2

1. $c = 31.4$ ft; $a = 78.5$ sq ft

2. $c \approx 13.2$ in.; $a \approx 13.8$ sq in.

3. $r \approx 2.7$ cm; $d \approx 5.4$ cm

4. $r \approx 3.4$ m; $d \approx 6.8$ m

5. 18.8 sq m   6. Answers will vary.

7. Answers will vary.   8. Answers will vary.

9. Answers will vary.

10. Sphere; there are no faces.

Short Quiz 12.4

1. 240 square units   2. 226.08 square units

3. See students' sketches; side should be labeled 5 centimeters.

4. 360 cubic units   **5.** $x = 12$ cm

6. See students' sketches.

7. 220 cubes   **8.** 238 square inches

■ **Mid-Chapter Test, 12-A**

1. $A = 254.34$ square feet; $C = 56.52$ feet

2. 50.24 square centimeters

3. 229.2 square inches   **4.** 43.96 square yards

5. Check students' sketches.

6. 6 feet   **7.** Check students' sketches.

■ **Mid-Chapter Test, 12-B**

1. $A = 379.94$ square yards; $C = 69.08$ yards

2. 39.25 sq ft   **3.** 437.76 sq m

4. 395.64 sq in.   **5.** Check students' sketches.

6. 4.5 feet   **7.** Check students' sketches.

■ **Short Quiz 12.6**

1. 401.92 cubic meters

2. Radius = 3 centimeters

3. Height = 2 millimeters

4. 64 cubic centimeters

5. 235.5 cubic feet

6. Check students' sketches. For a cylinder, the volume is 1130.4 cubic centimeters. For a cone, the volume is 376.8 cubic centimeters.

■ **Short Quiz 12.8**

1. 1436.03 cubic inches

2. 2224.97 cubic centimeters

3. About 2 feet   **4.** 6 inches

5. Check students' sketches and dimensions.

6. Check students' sketches and dimensions.

7. $x = 21.9$ feet; $y = 11.7$ feet

8. Cube or sphere

9. Answers will vary. In general, the scale factor of two dimensions in the dissimilar figures are different.

■ **Chapter Test, 12-A**

1. Circumference = 94.2 inches; area = 706.5 square inches

2. Circumference = 31.4 centimeters; area = 78.5 square centimeters

3. Cylinder   **4.** 9.42 feet

5. 7.07 square feet   **6.** 51.81 square feet

7. 28.26 cubic feet   **8.** Rectangular prism

9. 6 faces, 12 edges, 8 vertices   **10.** 264 m²

11. Find the area of the square base and multiply it by 2. Find the area of a rectangular side and multiply it by 4. Then, add these two numbers together.

12. 216 square feet   **13.** 150 square yards

14. 378 square centimeters

15. 140 cubic decimeters   **16.** 56 cubic feet

17. 392.5 cubic centimeters

18. ≈ 267.9 cubic inches

19. No   **20.** $x = 10$ **21.** $x = 12$

22. $x = 2.5$ **23.** 226.08 cm²

24. 251.2 cm³   **25.** 1:2 or $\frac{1}{2}$

26. 56.52 cm²   **27.** 31.4 cm³

■ **Chapter Test, 12-B**

1. b   **2.** a   **3.** c   **4.** b   **5.** c

6. d   **7.** d   **8.** b   **9.** a   **10.** a

11. d   **12.** d   **13.** c   **14.** a   **15.** c

16. c   **17.** d   **18.** b   **19.** a   **20.** c

21. b   **22.** b   **23.** c   **24.** d   **25.** a

■ **Chapter Test, 12-C**

1. Circumference = 46.47 m; area = 171.95 m²

2. 122.46 cm²   **3.** 69.08 ft   **4.** 11.0 ft

5. See students' sketches.   **6.** Hexagonal prism

7. 8   **8.** 18   **9.** 12   **10.** 60 ft²

11. 480 ft²   **12.** 900 ft³

13. Square pyramid

14. 246 ft²; one possible strategy: find the area of one base and multiply it by 2; add to it the product of 3 times the area of one rectangular side.

15. 77.98 m²; one possible strategy: find the area of each of the different rectangle pairs; double each area; then add all three to find the total.

16. $\approx$ 205 cm^2 (204.9792); one possible strategy: find the area of one base and multiply it by 2; add to it the product of the circumference of the base times the height.

17. See students' sketches; $h = 3.9$ cm, $d = 1.3$ cm

18. See students' sketches; 22.05 in. $\times$ 12.6 in. $\times$ 18.3 in.

19. Answers will vary; be sure that all dimensions reflect the same scale factor.

20. 65.42 m^3 21. 22.5 ft^3 22. 46.65 cm^3

23. 2,592,100 m^3 24. 3.5

25. 22,437.917 cm^3 26. 523.$\overline{3}$ cm^3

27. 42.875; it is approximately the cube of the scale factor of the radius/diameter (difference due to rounding).

■ **Cumulative Test, 7–12**

1. $1\frac{5}{24}$ 2. $\frac{9}{14}$ 3. $\frac{27}{50}$ 4. $\frac{5}{6}$

5. $x = 1\frac{1}{3}$ 6. $b = \frac{14}{15}$ 7. $e = \frac{3}{28}$

8. $h = \frac{1}{2}$ 9. 1.68   10. 0.46

11. 0.08   12. 1.51   13. $\frac{1}{2}$; 50%

14. $\frac{3}{8}$; 37.5%   15. $\frac{5}{9}$; 55.$\overline{5}$%

16. Pears: $\frac{7}{100}$; oranges: $\frac{1}{4}$; peaches: $\frac{1}{10}$; apples: $\frac{3}{10}$; bananas: $\frac{3}{25}$; grapes: $\frac{4}{25}$

17. Pears: 28; oranges: 100; peaches: 40; apples: 120; bananas: 48; grapes: 64

18. 5 students/1 table; rate   19. $\frac{11}{20}$; ratio

20. $x = 35$ 21. $y = 45$

22. $z = \frac{15}{4}$ 23. $x = \frac{4}{3}$

24. Increase; 44.9%   25. Decrease: 11.9%

26. $\frac{1}{10}$ 27. $\frac{2}{5}$ 28. $t = 13, -13$; rational

29. $r \approx 2.236, -2.236$; irrational

30. $e = 7, -7$; rational   31. $c \approx 17.49$

32. $b \approx 17.32$ 33. $a = 15$

34. $m > 4$;

35. $s \geq 6$;

36. $k \leq 1$;

37. $c < -52$;

38. $f < 8$;

39. No; $6^2 + 16^2 \neq 26^2$ (or $6 + 16$ is not greater than 26)

40. No; $8 + 15$ is not greater than 23.

41. Yes; $6 + 7 > 8$

42. $\overleftrightarrow{SI}, \overleftrightarrow{SH}, \overleftrightarrow{UH}$ or $\overleftrightarrow{IU}$ (answer should include 2)

43. $\overline{RK}, \overline{RO}, \overline{RU}, \overline{RT}, \overline{RE}$, or $\overline{RL}$ (answer should include 5)

44. $\overrightarrow{LR}$ or $\overrightarrow{LK}$ 45. $\angle ERU, LRT, ERT$

46. One possible answer: $\angle LEN$ and $\angle SER$; $\angle ORK$, and $\angle URE$

47. $\angle SER, LEN, URK, ORE$

48. One possible answer: $\angle ISF$ and SUT, and $\angle LES$, and ERU

49. One possible answer: $\angle TUS$ and URE, $\angle FSI$ and SEL

50. Check students' drawings; right

51. Check students' drawings; acute

52. Check students' drawings; straight

53. Check students' drawings; obtuse

54. Check students' responses on the drawing (two lines of symmetry); rotational symmetry at 180°.

55. Obtuse scalene triangle; $x = 38°$

56. Trapezoid; $x = 40°$ 57. Pentagon; 122°

58. $\angle X, \angle Y, \angle W$ 59. $\angle F, \angle E, \angle D$

60. Perimeter: $\approx$ 35.21 inches; area: 48 square inches

61. Perimeter: $\approx$ 29.62 yards; area: 52.5 square yards

62. Yes

63. They are equidistant from line m.

64. Circumference: 43.96 feet; area: 153.86 square feet

65. Radius: 4 inches; diameter: 8 inches

66. Check students' sketches and scale factors: given parallelogram has a perimeter of 45 inches.

67. Area of given figure: 90 square inches; see students' sketches and calculations

68. Surface area: 366 square centimeters; volume: 351 cubic centimeters

69. Surface area: 37.68 centimeters volume: 15.7 cubic centimeters

70. 108 cubic feet   71. $\approx$ 4.18 cubic feet

72.

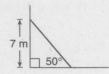

7 m   50°

73. 9.1 meters

■ **Short Quiz 13.2**

1. Yes   **2.** No (0, 2)

3. No (1, 5)   **4.** Yes

5.

x	-3	-2	-1	0	1	2	3
y	-5	-3	-1	1	3	5	7

6. $P = 2.2K$ **7.** 50 kg

8. and 9.

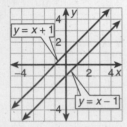

10. Yes; they have the same slope, therefore they will never meet.

■ **Short Quiz 13.4**

1.   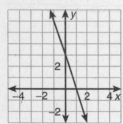   **2.**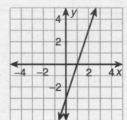

3.

x	-3	-2	-1	0	1	2	3
y	-12	-9	-6	-3	0	3	6

4. 2   **5.** $-\frac{1}{2}$

■ **Mid-Chapter Test, 13-A**

1. 16 ounces   **2.** 355 milliliters

3. A liter, 32 ounces = 947.2 milliliters

4. No

5.

x	-2	-1	0	1	2	5
y	$\frac{14}{3}$	4	$\frac{10}{3}$	$\frac{8}{3}$	2	0

6.

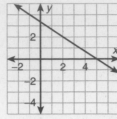

7. x-intercept: 5; y-intercept: $\frac{10}{3}$

8. $\frac{3}{5}$ **9.** $-\frac{1}{7}$

10.

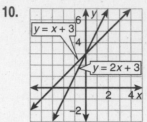

$y = x + 3$ $y = 2x + 3$

11. No; their slopes are different so there will be a point of intersection.

■ **Mid-Chapter Test, 13-B**

1. 8 feet   **2.** 5 meters

3. A mile; 5280 feet ≈ 1610 meters > 1000 meters

4. Yes

5.

x	-2	-1	0	1	2	4
y	18	15	12	9	6	0

6.

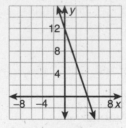

7. x-intercept: 4; y-intercept: 12

8. $\frac{3}{2}$ **9.** -1

10.

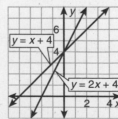

$y = x + 4$ $y = 2x + 4$

11. No; their slopes are different so there will be a point of intersection.

■ **Short Quiz 13.6**

1. Slope $= 2$; y-intercept $= 5$
See students' graphs.

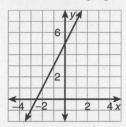

2. Slope $= \frac{1}{3}$;
y-intercept $= -2$
See students' graphs.

3. $y = -\frac{1}{2}x + 5$

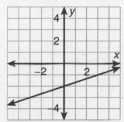

4. See students' graphs. **5.** $0.2x + 0.5y = 3.0$

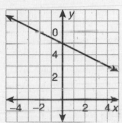

6. $(15, 0)$, $(0, 6)$

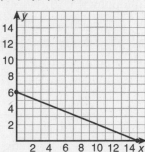

7. $(10, 3)$ is not a solution of the equation; 10 small beads and 3 large beads cost $3.50.

■ **Short Quiz 13.8**

1. Yes; $25 \geq 20$ **2.** Yes; $20 \geq 20$
3. Answers will vary.

4. Check students' graphs.

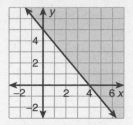

5. 6.71 **6.** $\left(-\frac{1}{2}, 0\right)$ **7.** 22.26

■ **Chapter Test, 13-A**

1. Yes **2.** No **3.** No **4.** Yes

5.

x	-3	-2	-1	0	1	2	3	4
y	-14	-12	-10	-8	-6	-4	-2	0

6. x-intercept: 4; y-intercept: -8

7. See students' graphs.

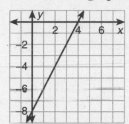

8. 2 **9.** -2 **10.** $\frac{1}{7}$ **11.** Horizontal
12. Vertical **13.** Slanted
14. See students' graphs.

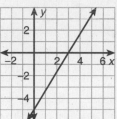

15. x-intercept: 5;
y-intercept: 5
See students' graphs.

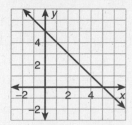

16. x-intercept: 2;
y-intercept: 4
See students' graphs.

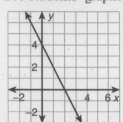

17. $y = -2x - 3$

18. See students' graphs. **19.** See students' graphs.

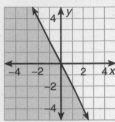 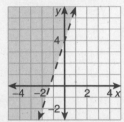

20. Midpoint: (4, 2.5); distance: 7.81
21. Midpoint: (5, 3); distance: 4.47
22. $x + 2y = 30$ **23.** 10
24. See students' graphs; solutions will vary.

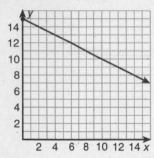

■ **Chapter Test, 13-B**

1. c	**2.** b	**3.** d	**4.** c	**5.** a	**6.** a
7. a	**8.** b	**9.** a	**10.** a	**11.** c	**12.** a
13. d	**14.** b	**15.** c	**16.** c	**17.** a	**18.** d
19. d	**20.** d	**21.** b	**22.** b		

■ **Chapter Test, 13-C**

1.

x	-3	-2	-1	0	1	2	3	6
y	6	$\frac{16}{3}$	$\frac{14}{3}$	4	$\frac{10}{3}$	$\frac{8}{3}$	2	0

2. x-intercept: 6; y-intercept: 4

3. See students' graphs.

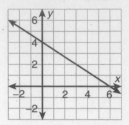

4. $-\frac{2}{3}$ **5.** $y = -\frac{2}{3}x + 4$ **6.** $-\frac{13}{6}$

7. Slope: $-\frac{1}{2}$; y-intercept: 3

8. Answers will vary. **9.** Answers will vary.

10. See students' graphs.

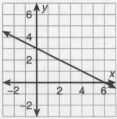

11. Graphs will vary but will have an equation of $y =$ _____.

12. Graphs will vary but will have an equation of $x =$ _____.

13. Graphs will vary but will have an equation of $y = mx + b$ or $y = mx$ or $y = x + b$.

14. $\frac{1}{2}x - 2y = 5$

15. (2, 6) and (5, 2); midpoint: (3.5, 4); distance: 5

16. $(-1, 3)$ and (7, 1); midpoint: (3, 2); distance: 8.25

17. See students' graphs. **18.** See students' graphs.

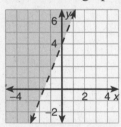 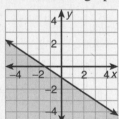

19. Answers will vary; about 440.

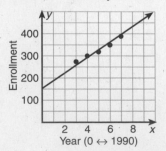

Year (0 ↔ 1990)

20. $3x + 5y \geq 35;\ y \geq -\frac{3}{5}x + 7$

21. See students' graphs.

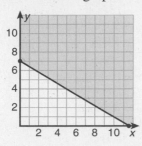

22. Answers will vary.

■ **Short Quiz 14.2**

1. 15.75, 13.5, 12   **2.** 4, 4, 4

3. 0, 1, 1, 1, 1, 1, 2, 2, 2, 2, 2, 2, 2, 3, 3, 3, 3, 4, 4, 5

4. 2.2, 2, 2   **5.** Answers will vary.

6. Answers will vary.

7.
9	1 1 3
8	1 1 1 2 2 3 3 4 5 6 6 7 7 7 9 9
7	2 6 7 8 8 8 9 9 9
6	7 8

9 | 1 represents 91.

8. See students' histograms.

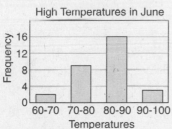

High Temperatures in June

■ **Short Quiz 14.4**

1. 4, 19   **2.** 7, 8.5, 12

3. 25%   **4.** 25%

5. Sum: $\begin{bmatrix} -12 & 8 \\ -9 & 3 \end{bmatrix}$

Difference: $\begin{bmatrix} 16 & -16 \\ -1 & 15 \end{bmatrix}$

6. Sum: $\begin{bmatrix} -2 & 11 & -7 \\ 0 & -2 & 12 \end{bmatrix}$

Difference: $\begin{bmatrix} -6 & 1 & 7 \\ 6 & -2 & -2 \end{bmatrix}$

7. Answers will vary.

■ **Mid-Chapter Test, 14-A**

1.
Girls		Boys
9 7 0	9	5 8
9 2 1 0	8	0 5 6 7 9
8 6 6	7	8 9
	6	6
2	5	7
	4	9

0 | 9 | 5 represents 90 and 95.

2. 82.5   **3.** about 81.8   **4.** 76

5. 49, 52, 57, 66, 76, 76, 78, 78, 79, 80, 80, 81, 82, 85, 86, 87, 89, 89, 90, 95, 97, 98, 99

6. 76   **7.** 81   **8.** 89

9.

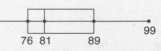

10. Answers will vary.

11. Sum: $\begin{bmatrix} 4 & 5 & -6 \\ 12 & -11 & -11 \end{bmatrix}$

Difference: $\begin{bmatrix} -12 & 9 & 4 \\ -6 & -5 & 11 \end{bmatrix}$

■ **Mid-Chapter Test, 14-B**

1.
Small		Large
8 1	9	1 2 4 9
7 5 5 2	8	7 9
7 6	7	6 8
3	6	3
9	5	
	4	9

1 | 9 | 1 represents 91 and 91.

2. 83.5   **3.** 81.8   **4.** 85

5. 49, 59, 63, 63, 76, 76, 77, 78, 82, 85, 85, 87, 87, 89, 91, 91, 92, 94, 98, 99

6. 76   **7.** 85   **8.** 91

9.

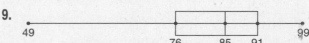

10. Answers will vary.

11. Sum: $\begin{bmatrix} -10 & 1 & 0 \\ 10 & 12 & -8 \end{bmatrix}$

Difference: $\begin{bmatrix} 6 & -9 & -6 \\ -10 & -2 & -4 \end{bmatrix}$

Short Quiz 14.6

1. Answers will vary, but polynomial will have exactly 1 term.

2. Answers will vary, but polynomial will have exactly 2 terms.

3. Answers will vary, but polynomial will have exactly 3 terms.

4. Check students' sketches.

5. $m^2 + 6m - 3$ 6. $4x^3 + 4x$

7. $5x^2 - 5x + 5$ 8. $-y^2 - 5y + 7$

9. $10b^3 - 5b^2 - 2b + 15$

10. $3c^3 - 9c^2 - 2c - 6$

Short Quiz 14.8

1. $12d^3 + 6d^2 - 3d$ 2. $-10f^4 + 20f^3 - 25f^2$

3. $-5w^5 + 4w^4 - 3w^3 + 2w^2$

4. $(x + 1)(2x - 1)$

5. $2(x + 1)(2x - 1) + 2(x)(2x - 1) + 2(x)(x + 1)$

6. $(x)(x + 1)(2x - 1)$

7. Answers will vary.

x	1	2	3	4	5
Volume	2	18	60	140	270

8. $2x^2 + 16x + 30$ 9. $12x^2 + 52x + 56$

10. $\frac{1}{2}(2x + 5)(3x + 7)$ 11. 165

Chapter Test, 14-A

1. 14.9   2. 13   3. 13, 9, 6

4. Answers will vary.

5.
```
2 | 0 3 4 6
1 | 3 3
0 | 6 6 9 9
```
2|0 represents 20.

6. 9, 13, 23

7. Histograms will vary. Check students' work. Accept all reasonable histograms.

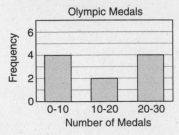

8. Group 1: 12.08; Group 2: 16

9. 2.5, 12, 19

10. 6, 18.5, 22.5

11.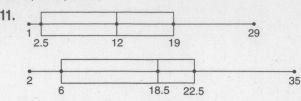

12. 2, 4, 6   13. 29.5, 38, 51.5

14. $\begin{bmatrix} 7 & 2 \\ -2 & 1 \end{bmatrix}$ 15. $\begin{bmatrix} 2 & 15 & -6 \\ 1 & 0 & 0 \\ 3 & -2 & -8 \end{bmatrix}$

16. $-8x^2 + 9x$; binomial

17. $-4x^2 + 8x + 3$; trinomial

18. x^2 19. $x^3 - 7x^2 + 14x - 16$

20. $12k^2 + 28$ 21. $40x^4 + 45x^3 - 20x^2$

22. $18n^2 + 19n + 5$ 23. $2(2x + 3) + 2(4x + 2)$

24. $(2x + 3)(4x + 2)$

25. Answers will vary.

x	1	2	3	4	5
Perimeter	22	34	46	58	70
Area	30	70	126	198	286

Chapter Test, 14-B

1. c   2. a   3. a   4. d   5. b   6. c

7. a   8. c   9. c   10. d   11. b   12. a

13. b   14. b   15. a   16. d   17. c   18. c

19. b   20. c   21. b   22. d   23. a

Chapter Test, 14-C

1. $302.5 million   2. $299.4 million

3. No mode   4. Answers will vary.

5. Stem-and-leaf plots will vary. Check students' work. Accept all reasonable plots. One possible solution:

Movie Grosses
(in millions)

24	24
25	12
26	00 37
28	58
31	29
32	27 97
35	71
39	98

24|24 represents 242.4.

6. Histograms will vary. Check students' work. Accept all reasonable histograms.

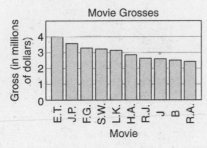

7.

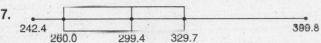

242.4   260.0   299.4   329.7   399.8

8. 68, 79, 82

9.

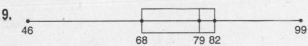

46   68   79 82   99

10. 4.1

11. 3; there would be 11 items whose sum is 44.

12. 3.5

13. One possible answer: 4; then the median item would be 4.

14. 4

15.

First Semester		Second Semester
1	9	0 1 3 5 5
7 6 5 3 1	8	2 8 9
8 2 0	7	6 8
4	6	

1|9|0 represents 91 and 90.

16. Answers will vary, but students may note that second semester's grades were higher.

17. $\begin{bmatrix} -1 & 0 & 7 \\ 6 & 6 & -12 \end{bmatrix}$ **18.** $\begin{bmatrix} 7 & 1 & -8 \\ 7 & -5 & 0 \\ 10 & 2 & 0 \end{bmatrix}$

19. Answers will vary.

20. $-3x^2$; monomial   **21.** $x + 5$; binomial

22. Answers will vary. Make sure students' trinomials have 3 terms.

23. $6x^4 + 3x^3 - 5x^2 - 4$

24. $-17x^4 + 4x^3 + 8x^2 - 8x - 2$

25. $32x^5 - 24x^3 + 56x^2$

26. $16x^2 + 40x + 25$